California Natural History Guides: 34

Native Shrubs of the Sierra Nevada

by
John Hunter Thomas
Stanford University and California
Academy of Sciences
and
Dennis R. Parnell
California State University,
Hayward

UNIVERSITY OF CALIFORNIA PRESS
BERKELEY • LOS ANGELES • LONDON

University of California Press
Berkeley and Los Angeles, California

University of California Press, Ltd.
London, England

ISBN: 0-520-02738-8 (clothbound)
ISBN: 0-520-02538-5 (paperbound)
Library of Congress Catalog Card Number: 73-86445
Printed in the United States of America

2 3 4 5 6 7 8 9

CONTENTS

Cypress Family
Mormon Tea Family
Agave Family
Willow Family
Bayberry Family
Birch Family
Oak Family
Elm Family
Birthwort Family
Goose-foot Family
Buttercup Family
Barberry Family
Calycanthus Family
Poppy Family
Caper Family
Saxifrage Family
Rose Family
Pea Family
Rue Family
Sumac Family
Staff Tree Family

Bladder-nut Family
Maple Family
Horse-chestnut or Buckeye
    Family
Buckthorn Family
Grape Family
Mallow Family
Cacao Family
Cactus Family
Oleaster Family
Dogwood Family
Silk Tassel Family
Heath Family
Styrax Family
Olive Family
Phlox Family
Waterleaf Family
Mint Family
Figwort Family
Madder Family
Honeysuckle Family
Sunflower Family

Acknowledgments

The signed line drawings are the work of Mrs. Jeanne R. Janish and were made expressly for this book. Unsigned drawings are the work of Mrs. Katherine K. Brown and are copyrighted by John H. Thomas and are used here by his permission.

The colored illustrations are from photographs by John H. Thomas, Roxana S. Ferris, and from the Dudley Herbarium and the California Academy of Sciences slide collections. A number of these are the work of Charles S.Webber.

Many thanks are due to Susan D. Thomas who has helped in many ways.

Cover photographs

Lower left: Chamise (*Adenostoma fasciculatum*) Photograph by Arthur C. Smith

Lower right: Mountain Pride (*Penstemon newberryi*) Photograph by Arthur C. Smith

Upper left: California Fremontia (*Fremontodendron californicum*) Photograph by Robert Ornduff

Upper right: Redbud (*Cercis occidentalis*) Photograph by Arthur C. Smith

INTRODUCTION

The area covered in this book, the Sierra Nevada, extends for nearly 400 miles from south of Mount Lassen, an area of active vulcanism, to Tehachapi Pass in the south and is one of the great mountain ranges of the world. The highest peak in the Sierra Nevada is Mount Whitney, 14,495 feet in elevation. The general directional axis of the range is southeast-northwest, and it is from 50 to 75 miles wide. The western slope rises gradually from the Central Valley of California to the crests, a distance of about 50 miles. This slope is dissected by numerous, generally westwardly oriented valleys, including Yosemite, many of which offer spectacular scenery. These have been formed through geological time by the erosive actions of rivers and glaciers as the Sierra Nevada has gradually been uplifted.

The eastern slope is much more precipitous. In many places the elevation drops from 12,000 to 5,000 feet in a linear distance of less than 15 miles. This slope is also dissected by deep valleys which are also the work of streams and glaciers. Some of the most spectacular scenery in the Sierra Nevada is along the crests of the range.

A few small glaciers persist in the Sierra Nevada, and it is not uncommon to find snow banks on north-facing slopes at high elevations in August.

Within the Sierra Nevada there are a number of life zones in the sense of Merriam. The lowest of these is the Upper Sonoran which extends along the length of the Sierra Nevada to the west and occurs in a few places to the east. Above this is the Transition Zone, occupying strips on both sides of the mountains. Above this are progressively the Canadian, Hudsonian, and Arctic-Alpine zones. The life zone concept is useful in a general way to describe very large regions, but does not offer enough "resolution" or sufficient categories to deal with the many plant communities present in the region.

Hence we are using the system of plant communities proposed by Munz and Keck (Munz, 1973). Some shrubs occur in only one community, others in two or more. Although a particular kind of plant may look very similar in a wide range of communities, it can occur in this wide range of communities because it has developed a series of ecological strains or ecotypes, each suited to slightly different environmental conditions.

Plant communities, and for that matter any biotic community, are not "things" in and of themselves. They are assemblages of plants that have mutually compatible tolerances to the main environmental factors of light, temperature, moisture, and soil.

The plant communities in the Sierra Nevada, together with some of their more conspicuous trees and shrubs, are listed below:

Alpine Fell Fields: Above timberline and usually above 10,500 feet. Very few shrubs, but occasionally *Haplopappus, Penstemon,* and *Salix.*

Chaparral: Extensive along the western slopes of the Sierra Nevada, usually below the Yellow Pine forest and rarely above about 4000 feet elevation. Conspicuous species are all shrubs, including: *Adenostoma fasciculatum, Arctostaphylos* (many species), *Ceanothus* (many species), *Cercocarpus betuloides, Eriodictyon californica, Pickeringia montana, Photinia arbutifolia, Quercus* (several species), and *Rhamnus* (several species).

Foothills Woodland: Foothills of the Sierra Nevada on western slopes, usually below 3000 feet. Prominent trees include: *Aesculus californica, Pinus sabiniana, Quercus agrifolia, Quercus chrysolepis, Quercus douglasii,* and *Umbellularia californica.* Among the conspicuous shrubs are: *Aesculus californica, Ceanothus* (several species), *Cercis occidentalis, Ribes* (several species), and *Rhamnus.*

Lodgepole Pine Forest: In the Sierra Nevada usually from about 8000 to 9500 feet, best developed in the central part of the mountain range. Lodgepole Pine, *Pinus murrayana* is the most conspicuous tree species.

Piñon-Juniper Woodland: Along the east base of the Sierra Nevada from about 4000 to 8000 feet. Dominants include: *Artemisia tridentata, Cercocarpus ledifolius, Juniperus* (several species), *Pinus monophylla,* and *Purshia tridentata.*

Red Fir Forest: Usually above 6000 and below 9000 feet. Important woody species are: *Abies magnifica, Castanopsis sempervirens, Ceanothus* (several kinds), *Pinus jeffreyi, Pinus monticola, Pinus murrayana,* and *Populus tremuloides.*

Sagebrush Scrub: Along the eastern base of the Sierra Nevada between about 3000 and 7500 feet. Shrubs include:*Artemisia cana, Artemisia tridentata, Chrysothamnus nauseosus, Purshia tridentata,* and *Tetradymia spinosa.*

Subalpine Forest: Usually from about 9500 to 11,000 feet. Prominent trees include: *Pinus albicaulis, Pinus murrayana,* and *Tsuga mertensiana.* Common shrubs include: *Cassiope mertensisna, Phyllodoce breweri, Ribes cereum,* and *Salix angelorum.*

Yellow Pine Forest: Western slopes from about 2000 to 6500 feet, but more usually from 3000 to 5000 feet. The most characteristic tree species are: *Libocedrus decurrens, Pinus ponderosa,* and *Quercus kelloggii.* Common shrubs include: *Actostaphylos* (several kinds), *Ceanothus* (several kinds), and *Chamaebatia foliolosa.*

About Shrubs And Their Names

It is difficult to list a group of characters which will precisely define a shrub. The numerous distinctions between herbs and shrubs, as well as between shrubs and trees, are somewhat arbitrary. Generally speaking, however, shrubs are defined as plants with one or several woody stems, with a maximum height of 10–15 feet. Certain shrubs may exceed these limits, for example, some willows and dogwoods.

Some trees have shrubby forms, a good example being the oaks, *Quercus.* These shrubby forms, a few of which are mentioned in this book, are the result of environmental conditions in the particular location in which the plants are growing (pl. 1, *a*). Had these plants grown in a more suitable environment, their habit would be that of a tree. On the other hand, a true shrub would maintain its habit, regardless of the conditions under which it is grown, with few exceptions.

Again, it is possible to confuse some herbs that have woody-based stems, with shrubs. However, herb stems die back every year, and therefore such plants are not considered to be shrubs.

It is easy then, to recognize a tree and equally easy to recognize a shrub. But distinguishing between a treelike shrub or a shrublike tree may cause trouble.

The common names of shrubs arise as a result of popular usage in a particular place or region. Therefore many shrubs may have more than one common name associated with them. So far there is no standardization of common names of plants in the United States.

Each kind of plant, or species, does have one standardized botanical name associated with it. This is sometimes called the "scientific name" and is of Latin origin or latinized to conform to the rules of that language. Species names are given to groups of populations of plants which appear to be similar to one another and genetically very closely related to each other. The factors which determine whether or not a group of populations "form" a species vary from group to group. For example, the criteria used for determining a species of rose may differ from the criteria used for species of manzanita.

Often a subgroup of populations belonging to a single species is given a varietal name, for example, *Baccharis pilularis* var. *consanguinea.* Where appropriate and useful, such names are used in this book.

Based on similarities in form and structure and presumed evolutionary relationships, species are grouped into genera (plural of genus). For example, all roses are in the genus *Rosa* and all manzanitas are in the genus *Arctostaphylos.* Genera are grouped into families, the genera *Adenostoma, Potentilla,* and *Rosa* (and others) into the Rosaceae.

ACTIVITIES

There are many ways to become acquainted with the Sierran shrubs. The best is to explore the Sierra Nevada in the course of several trips. Visits should cover different times of the year. As a result of such trips you will not only become familiar with the distribution and relative abundance of the various species, but also you will be able to observe when the Sierran shrubs come into flower.

You should not limit your study of shrubs to those that are found along roadsides. Many Sierran shrubs are found only along trails away from roadsides and other disturbed sites. Another advantage of examining the flora along the trails is that you will have a better indication of what natural habitats for these shrubs are like.

To identify shrubs you will need only a hand lens in addition to this book. Plant collecting is not permitted in national and state parks or monuments. It is possible to collect on private land, but only with permission of the owner. Also, local regulations may apply. Therefore, only the smallest amount of material (e. g., a single leaf or flower) should be taken for identification purposes. A good conservation-minded way of building up a reference collection of Sierran shrubs is by photographs or slides.

Finally, if it is not possible to go to the Sierras it is still possible to see Sierra Nevada shrubs. In both northern and southern California there are several botanical gardens where native Sierran shrubs are growing. For example in northern California there are both Tilden Park and the University of California Botanical Gardens in Berkeley; in San Francisco one may visit Strybing Arboretum in Golden Gate Park. In southern California visits to Rancho Santa Ana Botanic Garden in Claremont, the Santa Barbara Botanic Garden in Santa Barbara, as well as the Los Angeles County Arboretum in Arcadia, will be profitable. All of these gardens can supply information on how Sierran shrubs may be grown and where they may be purchased. (It should be emphasized that shrubs may only be dug up on pri-

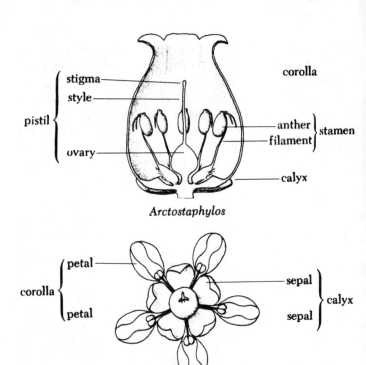

Arctostaphylos

Ceanothus

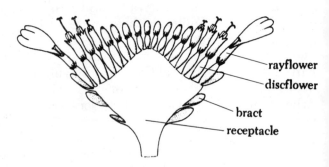

Compositae

Parts of flowers (from R. S. Ferris, *Native Shrubs of the San Francisco Bay Region*, University of California Press, 1968)

vate land and only with written permission of the owner. Also, shrubs usually don't survive transplanting.)

In addition, most of the above gardens are associated
with an herbarium which houses collections of Sierran
plants. Often when proper advance arrangements have
been made the personnel of these herbaria will assist
you in identifying specimens with which you have had
particular difficulty.

USE OF THIS BOOK

The information about various groups of plants is arranged in the form of tables and lists. Following are lists
of vines, plants with spines, plants with leaves with
more than one leaflet, plants with opposite leaves, plants
which lack petals, plants with flowers in catkins, plants
with irregular flowers, and finally lists arranged by
flower color.

In families and genera with a number of species, other tables, giving some of the important identifying
characters, are provided. In some cases it may be im-
possible to distinguish between two closely related
species unless a necessary part of the plant, such as the
mature fruit, is available. Another way of identification
is, of course, just to look at the pictures. Many of the
common shrubs can be identified easily in this way. To
learn about plants, it will be necessary to learn botanical terminology. At the beginning, everything from ask-
ing people to tell you about plants to looking at pictures is helpful.

Vines and vinelike shrubs

Shrubs with spines, spinescent branches, or spiny leaves

[11]

Shrubs with leaves having more than one distinct leaflet, or
leaves very finely divided, almost fernlike

Shrubs having leaves opposite each other on the stem or
more than two at a node

Shrubs with flowers lacking petals

* The sepals are petal-like.

Shrubs with flower small, unisexual, usually lacking at least petals, arranged in elongate clusters, known as catkins

Shrubs with irregular flowers

* Represents a calyx.

[13]

Shrubs with white or whitish flowers

* These represent sepals.

Shrubs with yellow flowers

Shrubs with pinkish to pale rose flowers

* Calyx resembles a corolla.

DESCRIPTIVE LIST OF THE NATIVE SHRUBS OF THE SIERRA NEVADA

The arrangement of plant families follows a well-known botanical system called the Engler and Prantl sequence. Within the families, the genera and species are arranged alphabetically.

Measurements of plants and plant parts are given in the metric system. A metric rule is printed on the inside of the front cover, and the following approximate equivalents may be helpful:

1 in. = 2.5 cm (centimeter)   10 mm (millimeter) = 1 cm

4 in. = 10 cm   10 cm = 1 dm (decimeter)

39 in. = 1 m (meter)   10 dm = 1 m

Altitudes in the Sierra Nevada are given in feet.

CYPRESS FAMILY (CUPRESSACEAE)

Three species of junipers occur in the Sierra Nevada.

Dwarf Juniper (*Juniperus communis*) (fig. 1). A low prostrate shrub, occasionally to 1 m tall, with sharp, rigid leaves, 6–12 mm long; stamens are in small conelike structures, 3–6 mm long; one to three seeds are borne in smooth, globose berries, which are 7–9 mm across

Fig. 1. *Juniperus communis*

and light blue at maturity. This distinctive conifer occurs in the Sierra Nevada from Mono Pass north in the Red Fir and Lodgepole Pine Forests.

California Juniper (*Juniperus californica*). A dense shrub from 1–4 m tall or sometimes a small tree, to about 10 m tall, with overlapping scalelike leaves about 3–4 mm long, arranged in 2's and having small glandular pits on the back; stamens are in small conelike structures; one to two seeds are enclosed within globose berrylike cones, 12–18 mm long, which are blue at first, but turn reddish-brown as the waxy coating is lost. California Juniper occurs mainly in the Foothill Woodland on the western slopes of the Sierra Nevada from Mariposa to Kern County.

Utah Juniper (*Juniperus osteosperma*). Occurs in the vicinity of Bridgeport in shrublike form. Similar to California Juniper but leaves arranged in 3's.

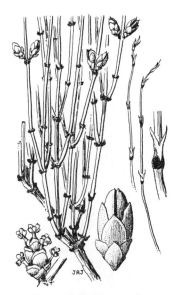

Fig. 2. *Ephedra viridis*

MORMON TEA FAMILY (EPHEDRACEAE)

Green Ephedra (*Ephedra viridis*) (fig. 2). A much-branched shrub from 0.5–1.5 m tall, with jointed bright green or yellow-green slender stems arising from a thick trunk and two scalelike leaves, less than 5 mm long, at each node. In this species the sexes are found on different plants, the female having at the leaf nodes scaly, seed-bearing cones, each containing two seeds; the male plants carrying the pollen-bearing cones. Both types of the plant usually grow close together. This species is found along the eastern slopes of the Sierra Nevada below about 7500 feet.

AGAVE FAMILY (AGAVACEAE)

Spanish Bayonet (*Yucca whipplei*) (fig. 3). A very distinctive plant with long, narrow, swordlike leaves to 1 m long, in a basal clump from which an erect flowering stalk to 2.5 m arises; flowers are numerous, 2.5–3.5 cm long and creamy-white in color; fruits are large cap-

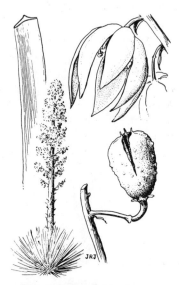

Fig. 3. *Yucca whipplei*

sules, about 3–4 cm long. It occurs in the southern Sierra Nevada from the region of the Kings River south to the Joshua Tree and Piñon-juniper Woodlands.

Spanish Bayonet is also known by several other names, among which are Our Lord's Candle, Quixote Plant, and Chaparral Yucca.

WILLOW FAMILY (SALICACEAE)

Willows occur in various forms. Some are trees; others are shrubs; and still others, the dwarf willows of high altitudes or latitudes, are low plants with the main stems underground and only small twigs, leaves, and catkins above ground. Willow flowers are very much reduced and lack both petals and sepals. Female flowers are on different trees or shrubs than the male; in both cases, the flowers are arranged in catkins. Female flowers consist of a single ovary which contains many minute, hairy seeds; these are shed from the mature capsulelike fruit and produce large amounts of a cottony fluff. Male flowers have 1–10 stamens. Willow leaves are usually elongate and alternate.

Willows are wind pollinated and the catkins appear before the leaves. The different kinds of willows hybridize and may be hard to identify. In any case, it may be necessary to have specimens of both male and female plants with catkins, and specimens of both kinds of plants with leaves in order to be sure of the identity of a willow. Most willows occur in moist places, at the edges of streams and lakes and in boggy ground.

Tables 1 to 5 will aid in identification.

I. Willows forming dense mats; the leaves entire (Table 1).

II. Erect willows with toothed leaves; leaves persistently hairy beneath (Table 2).

III. Erect willows with toothed leaves; leaves lacking hairs when mature (Table 3).

IV. Erect willows with entire leaves; leaves persistently hairy below (Table 4).

V. Erect willows with entire leaves; leaves lacking hairs beneath when mature (Table 5).

TABLE 1. Willows forming dense mats, leaves entire
(Stamens 2 per flower; filaments lack hairs.)

	Leaf length	Catkin length	Capsule length
S. angelorum	2–4 cm	1–6 cm	5–7 mm
S. nivalis	0.7–1.2 cm	1 cm	2.5–3.5 mm

TABLE 2. Erect willows with toothed leaves; leaves persistently hairy beneath
(Two stamens per flower; filaments hairy at base.)

	Height	Leaf length and width	Capsule	Capsule length	Style length	Stigma length
S. eastwoodiae	0.5–2 m	4–6 x 1–2 cm	with woolly hairs	5–6.5 mm	1–1.5 mm	0.5 mm
S. exigua	2–3 m	5–12 x 0.2–0.8 cm	lacking hairs	5 mm	lacking	0.5 mm
S. hindsiana	3–7 m	4–8 x 0.3–0.6 cm	with woolly hairs	5–6 mm	0.5 mm	1 mm
S. melanopsis	3–5 m	4–7 x 0.6–1.4 cm	with a few hairs	4–5 mm	lacking	0.2 mm

Table 3. Erect willows with toothed leaves; leaves lacking hairs when mature
(Twig color is given with statement of range.)

	Height	Leaf length and width	Catkin scale duration	Catkin scale color	Capsule	Capsule length	Stamen number	Stamen
S. caudata	1–4 m	6–12 x 1.5–3 cm	deciduous	yellow	without hairs	5–7 mm	3–9	hairy
S. lemmonii	1–5 m	3–10 x 0.8–2 cm	persistent	dark	with silky hairs	6–9 mm	2	hairy
S. ligulifolia	1–5 m	5–10 x 1–2 cm	persistent	dark	without hairs	4.5–5 mm	2	not hairy
S. lutea	2–5 m	4–10 x 1.5–4 cm	deciduous	yellow	without hairs	4–5 mm	2	not hairy
S. mackenziana	2–6 m	6–12 x 2–3 cm	persistent	dark	without hairs	4.5–5.5 mm	2	not hairy
S. melanopsis	3–5 m	4–7 x 0.6–1.4 cm	deciduous	yellow	without hairs	4–5 mm	2	hairy
S. pseudocordata	1–4 m	3–8 x 1–3 cm	persistent	dark	without hairs	4–6 mm	2	not hairy

(Two stamens per flower.)

	Height	Leaf length and width	Capsule length	Capsule	Style length	Stigma length	Stamens
S. drummondiana	1–4 m	3–7 x 1.5–2 cm	3.5–5 mm	with silky hairs	1–1.5 mm	0.2–0.5 mm	without hairs
S. exigua	2–4 m	5–12 x 0.2–0.8 cm	5 mm	without hairs	0	0.5 mm	hairy
S. geyeriana	3–5 m	2–7 x 0.6–1.5 cm	5–7 mm	with silky hairs	0.3 mm	0.3–0.5 mm	hairy
S. hindsiana	2–7 m	4–8 x 0.3–0.6 cm	5–6 mm	with woolly hairs	0.5 mm	1 mm	hairy
S. jepsonii	1–3 m	3–7 x 1–2 cm	4.5–5.5 mm	with densely silky hairs	0.7–1 mm	0.4 mm	without hairs
S. lasiolepis	2–10 m	6–10 x 1–2 cm	4.5–5 mm	without hairs	0.5 mm	0.3 mm	without hairs
S. orestra	1–3 m	4–6.5 x 1–2 cm	7–8.5 mm	with hairs	1 mm	0.2 mm	hairy
S. scouleriana	1–10 m	3–10 x 1.5–3 cm	7–9 mm	with woolly hairs	0.2–0.5 mm	0.5–1 mm	without hairs

TABLE 5. Erect willows with entire leaves; leaves lacking hairs beneath when mature
(Two stamens per flower.)

	Height	Leaf length and width	Capsule length	Capsule	Stamens
S. lasiolepis	2–10 m	6–10 x 1–2 cm	4–4.5 mm	without hairs	without hairs
S. lemmonii	1–5 m	3–10 x 0.8–2 cm	6–9 mm	with silky hairs	hairy
S. ligulifolia	1–5 m	5–10 x 1–2 cm	4.5–5 mm	without hairs	without hairs
S. lutea	2–5 m	4–10 x 1.5–4 cm	4–5 mm	without hairs	without hairs
S. planifolia	0.3–1.5 m	2–3.5 x 0.8–1.5 cm	5–7 mm	somewhat silky	without hairs

Alpine willow (*Salix angelorum*) (pl. 1, *b*). A low creeping willow which occurs in moist meadows. It ranges from Tulare County north, usually above 9000 feet in open areas in Subalpine Forests and Alpine Fell Fields.

Caudate Willow (*Salix caudata*). Has reddish-brown twigs. It occurs along streams and in moist meadows from 6000 to 8500 feet in the Red Fir and Lodgepole Pine Forests.

Drummond's Willow (*Salix drummondiana*). Occurs from Tulare to Inyo and Fresno counties, usually between 8500 and 9500 feet, in the Lodgepole Pine and Subalpine Forests in moist open areas.

Eastwood's Willow (*Salix eastwoodiae*). Occurs along stream banks and in meadows from Tulare County north from 7000 to 11,000 feet in coniferous forests.

Narrow-leaved Willow (*Salix exigua*) (pl. 1, *c*). Characterized by its linear leaves. It is common on the eastern slopes of the Sierra Nevada along streams below 8000 feet, and it occasionally occurs also on the western slopes in Kern County.

Geyer's Willow (*Salix geyeriana*). Occurs in montane forests from Tulare County north between about 5000 and 10,000 feet.

Jepson's Willow (*Salix jepsonii*). Ranges from Tulare County north between about 5000 and 10,000 feet in moist areas in several areas in several plant communities.

Sandbar Willow (*Salix hindsiana*). Common below about 3000 feet on sand bars, along ditches, and in other moist areas on the western slopes of the Sierra Nevada. This species is usually a tree.

Arroyo Willow (*Salix lasiolepsis*) (pl. 1, *d*). Very common along streams below 7000 feet in several plant communities. This species is usually a tree.

Lemmon's Willow (*Salix lemmonii*). Has yellowish twigs and ranges from Tulare County north between 5000 and 10,000 feet in the Sierran coniferous forests.

Ligulate Willow (*Salix ligulifolia*). Has yellowish twigs and occurs from Tulare County north between 3500 and 9500 feet in a number of plant communities.

Yellow Willow (*Salix lutea*). Has yellowish or brown twigs and occurs in the northern Sierra Nevada from Nevada County north.

MacKenzie's Willow (*Salix mackenziana*). Has dark brown twigs and occurs in the Red Fir forest from Tulare County north. This species is usually a tree.

Dusky Willow (*Salix melanopsis*). Has dark-colored twigs and occurs along stream banks below 8000 feet in several plant communities.

Snow Willow (*Salix nivalis*). Quite rare in the Sierra Nevada, being known only from several locations in Mono County between 10,000 and 12,000 feet.

Sierra Willow (*Salix orestra*). Occurs at high elevations, 8500 to 12,000 feet, from Tulare County north in meadows, along stream banks, and along the shores of lakes.

Mono Willow (*Salix planifolia*). Occurs at high elevations, 8000 to 12,500 feet, in the central Sierra Nevada in Fresno, Tuolumne, Inyo, and Mono counties.

[23]

Fig. 4. *Myrica hartwegii*

Firmleaf Willow (*Salix pseudocordata*). Has yellowish to brownish twigs and ranges from Tulare County north between 6000 and 11,000 feet along streams and lake banks.

Scouler's Willow (*Salix scouleriana*). A very common willow of stream banks and marshy areas below about 10,000 feet. Scouler's Willow is usually a tree.

Bayberry Family (Myricaceae)

Sierra Wax Myrtle (*Myrica hartwegii*) (fig. 4). A small deciduous shrub to about 2 m tall with slender, coarsely toothed leaves, 4–8 cm long. Male flowers occur in small catkins to 2 cm long; female flowers in catkins about 8 mm long in fruit. Both kinds of flowers lack speals and petals. Sierra Wax Myrtle occurs along stream banks, between 1000 and 5000 feet, mainly in the Yellow Pine Forest from Yuba to Fresno counties, hence it is endemic to the western slopes of the Sierra Nevada.

Fig. 5. *Betula occidentalis*

It is also known as Sierra Sweet Bay, although it bears no relationship to the California Bay (*Umbellularia californica*).

BIRCH FAMILY (BETULACEAE)

The birch family is represented in the Sierra Nevada by three genera. *Corylus* has leaves which are heart-shaped at the base; the other two genera do not. *Betula* has the female catkins borne singly; those of *Alnus* are in clusters. The female catkins of *Betula* come apart at maturity; those of *Alnus* do not, but fall as a conelike structure.

Mountain Alder (*Alnus tenuifolia*). A shrub or small tree with leaves from 2.5–10 cm long. Male catkins are borne in groups at the tips of the branchlets; female catkins are borne in groups at the tips of the branches and become woody at maturity. Mountain Alder grows at elevations below 8500 feet from Tulare County north in mixed coniferous forest.

Fig. 6. *Corylus californica*

Water Birch (*Betula occidentalis*) (fig. 5). A tall shrub or more usually a small tree. The oval leaves are 1–6 cm long, with strongly toothed margins. The sharp-scaled cylindrical male catkins may grow to 5 cm or more in length and are clustered at the tips of the branches. Female catkins with their characteristic scaly bracts are borne singly and are rarely over 2.5 cm long; the red stigmas are quite prominent when protruding from the green bracts. Water Birch is most frequent in wet or damp localities in mixed coniferous forest at elevations below about 8000 feet.

California Hazelnut (*Corylus californica*) (fig. 6). A freely spreading shrub which may grow to about 6 m in height with tooth-edged leaves 4–7 cm long. Female flowers are contained in small rounded scalelike clusters from which bright red stigmas protrude. Male flowers are in hanging catkins. The fruit is an edible nut about 1.5 cm in diameter. California Hazelnut is widespread

Fig. 7. *Castanopsis sempervirens*

in the Sierra Nevada from Tulare County north at elevations below about 7000 feet.

Oak Family (Fagaceae)

The three genera of the oak family which are native to California all have shrub species or shrubby forms of tree species in the Sierra Nevada. All are evergreen and have thick, leathery leaves. In *Castanopsis* the nuts are enclosed in a very spiny burlike structure; in *Lithocarpus* and *Quercus*, the fruit is easily recognized as an acorn. *Lithocarpus* has erect male catkins and leaves with conspicuous parallel veins, while *Quercus* has pendant male catkins and leaves with generally nonparallel veins.

Bush Chinquapin (*Castanopsis sempervirens*) (fig. 7). Can be identified easily by its unique burred fruit and round-tipped evergreen leaves. Leaves are gold colored beneath, due to dense feltlike hairs. Bush Chinq-

[27]

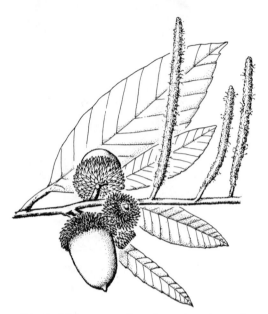

Fig. 8. *Lithocarpus densiflora var. echinoides*

uapin is 0.5–2.5 m high and often much wider across. It is found in the Sierra Nevada on rocky slopes, between about 4000 to 11,000 feet. The shrubby form of the Giant Chinquapin known as the Golden Chinquapin (*Castanopsis chrysophylla* var. *minor*) has pointed and somewhat folded leaves. It occurs infrequently on the western slopes of the Sierra Nevada in El Dorado County.

Tanbark Oak (*Lithocarpus densiflorus* var . *echinoides*) (fig. 8). The shrubby form of this oak occurs in the Yellow Pine and Red Fir Forests from Mariposa County north. It is 1–3 m tall. In this genus, unlike the true oaks, the acorn cup has stiff spreading scales. Its leaves are 2.5–7 cm long with prominent parallel lateral veins. The female flowers are at the bases of the clusters of erect male flowers.

Oaks (*Quercus*)

Oaks with their drooping catkins of male flowers and

Fig. 9. *Querous vaccinifolia*

characteristic acorns are a common element of the California landscape. Although a number of species, usually considered to be trees, have shrubby forms (*Quercus chrysolepsis, Q. garrayana, Q. kelloggii,* and *Q. wislizenii*: see *Native Trees of the Sierra Nevada* in the references), three distinct shrubs occur in the Sierra Nevada.

Scrub Oak (*Quercus dumosa*). Usually about 1–3 m tall. Leaves are flat, either with or without spiny margins, and generally devoid of any hairs on the upper surface, thus appearing rather shiny. Scrub Oak occurs in Chapparral and Foothills Woodland at lower elevations on the western slopes of the Sierra Nevada.

Leather Oak (*Quercus durata*) (pl. 1, *e*). The convex leaves are matted with short hairs on the upper surface. Leaf margins are entire or variously toothed. All records so far indicate that Leather Oak is restricted to serpentine soils, generally in Chaparral from Nevada to El Dorado counties.

[29]

Fig. 10. *Celtis douglasii*

Huckleberry Oak (*Quercus vaccinifolia*) (fig. 9). A low, widely spreading shrub, to about 1.5 m but usually less in height. Leaves are gray-green on the upper surface usually with entire margins; hairs are lacking on the upper leaf surface. Huckleberry Oak occurs in open, dry places from about 5000 to 10,000 feet from Fresno County north.

ELM FAMILY (ULMACEAE)

Douglas' Hackberry (*Celtis douglasii*) (fig. 10). A shrub or sometimes a small tree, usually 2–6 m tall. Leaves are pointedly egg-shaped, 3–6 cm long. Male and female flowers are small and occur on different parts of the same plant; both kinds of flowers lack petals. The fruit is a globose berry, orange-brown at maturity and about 6–8 mm long. This shrub occurs in isolated localities in the southern Sierra Nevada.

[30]

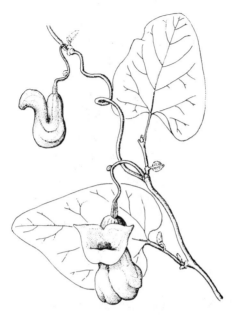

Fig. 11. *Aristolochia californica*

Birthwort Family (Aristolochiaceae)
California Pipe Vine or Dutchman's Pipe (*Aristolochia californica*) (fig. 11). A straggly woody vine, climbing over other vegetation to about 3 m in height. Leaves are narrowly heart shaped, 4–15 cm long, with a velvety texture on both surfaces. The conspicuous part of the flower is the calyx, which is pendant and purplish, about 2–4 cm long, and resembles a pipe. Petals are not present. The fruit, which is below the calyx, becomes a capsule 3–4 cm long and eventually opens to shed the 6 mm-long seeds.

This vine occurs from El Dorado County north in the Sierra Nevada foothills, usually below about 2000 feet, mainly along stream banks. It has also been collected in Fresno County.

Fig. 12. *Grayia spinosa*

GOOSEFOOT FAMILY (CHENOPODIACEAE)

Two members of the goosefoot family have shrubby members in the Sierra Nevada. In *Grayia* the leaves are flat, while in *Eurotia* the leaf margins are rolled under. Both species are nondescript looking plants, grayish and with inconspicuous flowers.

Winter Fat (*Eurotia lanata*) (pl. 1, *f*). A low erect shrub to 8 dm tall. The herbage is covered with a dense layer of whitish scales, interspersed with white hairs. Leaves, which are in bundles, are slender, elongate, 1.5–5 cm long. Flowers are small and inconspicuous; male and female flowers are on different plants. This is an important forage plant in an unfavorable season, hence the name. It occurs along the eastern slopes of the Sierra Nevada in dry areas at low elevations.

Fig. 13. *Clematis lasiantha*

Hopsage (*Grayia spinosa*) (fig. 12). A small, 3–10 dm tall, somewhat spinescent and much-branched shrub. Leaves are elongate, 1–3 cm long, and alternate. Flowers are small in elongate clusters, both sepals and petals lacking; male and female flowers are on different plants. Hopsage occurs occasionally on the eastern slopes of the Sierra Nevada in Pinon-Juniper Woodland, below 7500 feet and usually above 3000.

BUTTERCUP FAMILY (RANUNCULACEAE)

Two species of virgin's bowers are found in the Sierra Nevada. Both are vines with opposite leaves with leaflets. The flowers of virgin's bower are conspicuous, the female parts forming fuzzy headlike balls, 4–6 cm across, which remind one of Christmas ornaments out of season. Petals are lacking, but the sepals, white, 1–2.5 cm, and showy, substitute for the petals.

[33]

Fig. 14. *Berberis pumila*

Western Virgin's Bower (*Clematis ligusticifolia*). Leaves have 5–7 leaflets; sepals are about 1 cm long; many flowers occur in a cluster. This species occurs below about 7000 feet in moist places in several plant communities.

Chaparral Virgin's Bower (*Clematis lasiantha*) (fig. 13, 2, *a*). Leaves usually have only 3 leaflets. Flowers are in clusters of no more than three; sepals are 1.5–2.5 cm long. This species is most common in Chaparral, below about 4500 feet on the western slopes of the Sierra Nevada.

BARBERRY FAMILY (BERBERIDACEAE)

Two species of barberries occur in the Sierra Nevada. Both are small, woody subshrubs, under 0.5 m tall, with alternate leaves composed of several leaflets. The flowers are in elongate clusters and have six sepals and six yellow petals. The fruits are bluish-black berries, about 6 mm in diameter at maturity.

[34]

Fig. 15. *Calycanthus occidentalis*

Sonne's Barberry (*Berberis sonnei*). Leaflets are light green and glossy above. Sonne's Barberry is endemic in the eastern Sierra Nevada, primarily on rocky slopes in the coniferous forests between 6000 and 8000 feet.

Dwarf Barberry (*Berberis pumila*) (fig. 14). Leaflets are dullish gray-green. Dwarf Barberry occurs in the Yellow Pine Forest on the western slopes of the Sierra Nevada from Mariposa County north below 4000 feet.

CALYCANTHUS FAMILY (CALYCANTHACEAE)
Spice Bush (*Calycanthus occidentalis*) (fig. 15). A rounded shrub to 3 m tall, or rarely taller in favorable sites. Leaves are opposite, narrowly egg-shaped, 5–15 cm long, and fragrant when crushed, hence the name Spice Bush. Petals and sepals are purplish, 2–6 cm long, and look similar. This is one of the few Sierra shrubs with numerous stamens. The fruit, more correctly the

Fig. 16. *Dendromecon rigida*

floral tube is 3–4 cm long. Spice Bush, which is also known as Sweet Shrub or Western Sweet-Scented Shrub, occurs in moist slopes of the Sierra Nevada from Tulare County north in the Foothills Woodland and Yellow Pine Forest. A good place to see this shrub is in the vicinity of Arch Rock Ranger Station along the Merced Highway into Yosemite National Park.

POPPY FAMILY (PAPAVERACEAE)

Tree Poppy (*Dendromecon rigida*) (fig. 16). A distinctive shrub of Chaparral and dry ridges, generally 1–3 m tall, with stiff, whitish branches and thick elongate leaves 2.5–10 cm long. The three sepals fall from the flower, exposing four bright-yellow petals. Flowers are most common from April to June. The fruit is an elongate capsule, 5–10 cm long. Tree Poppy occurs below 6000 feet from Tulare County north on the western slopes of the Sierra Nevada.

Fig. 17. *Isomeris arborea*

CAPER FAMILY (CAPPARACEAE)

Bladder Pod (*Isomeris arborea,* or perhaps more correctly, *Cleome arborea*) (fig. 17). An ill-scented shrub to 2.5 m tall. The alternate leaves have three, narrow leaflets, 1–3 cm long. Flowers are in elongate clusters, with conspicuous yellow petals, 1–1.5 cm long. Mature fruits are bladdery, 2.5–3 cm long, and are on stalks 1–2 cm long, above the place where the sepals and petals were attached. This is largely a desert and arid-land shrub of much of the southwest, but it does ocur in drier areas adjoining the southern part of the Sierra Nevada.

SAXIFRAGE FAMILY (SAXIFRAGACEAE)

Four genera of this family occur in the Sierra Nevada. They are distinguished in Table 6.

Tree Anemone or Carpenteria (*Carpenteria californica*) (fig. 18). A spreading evergreen shrub, usually

TABLE 6. Genera in the Saxifrage Family

Genus	Leaf arrangement	Leaf duration	Leaf length	Spines	Petal length	Petal color
Carpenteria	opposite	persistent	5–8 cm	absent	2–3 cm	white
Jamesia	opposite	deciduous	1–2.5 cm	absent	0.6–0.7 cm	rose-pink
Philadelphus	opposite	deciduous	2–5 cm	absent	1–1.5 cm	white
Ribes	alternate	persistent or deciduous	1–8 cm	present or absent	0.1–1 cm	white to pink, or rose, also yellow

TABLE 7. Species in the Genus Ribes with Spines

	Height	Number nodal spines	Leaf width	Number leaf lobes	Degree of lobing	Length of free part of flower tube	Ovary and berry	Berry diameter	Berry color
R. amarum	1–2 m	3–5	2–3 m	3–5	shallow	5–6 mm	spiny or bristly	15–20 mm	usually purplish
R. divaricatum	1–3 m	1–3	2–5 cm	3–5	shallow	3–4 mm	no bristles	6–10 mm	purplish
R. lasianthum	1 m	1–3	1–2 cm	3–5	shallow	4 mm	no bristles	6–7 mm	red
R. menziesii	1–2 m	3	1.5–4 cm	3–5	shallow	2–3 mm	spiny or bristly	10–25 mm	usually reddish
R. montigenum	0.3–0.6 m	3–5	0.5–25 cm	5	very deep	0–0.5 mm	bristly	5 mm	red
R. quercetorum	0.6–1.6 m	1	1–2 cm	3–5	deep	2.5 mm	no bristles	7–8 mm	black
R. roezlii	0.5–1.2 m	1–3	1.2–2.5 cm	3–5	deep	6 mm	spiny and bristly	14–16 mm	purple
R. tularensis	0.5 m	3	2–5 cm	3	deep	4 mm	spiny or bristly	10 mm	purplish

TABLE 8. Species in the Genus *Ribes* without Spines

	Flower color	Flower tube length	Petal length	Sepal length	Leaf width	Berry color	Berry diameter	Pubescence
R. aureum	yellow	6–10 mm	2–3 mm	5–8 mm	1.5–5 cm	red-black	6–8 mm	lacking
R. cereum	whitish	6–8 mm	1 mm	1–1.5 mm	1–4 cm	red	6 mm	glandular
R. inebrians	whitish	6–8 mm	1 mm	1–1.5 mm	1–4 cm	red	6 mm	glandular
R. malvaceum	light rose	5–8 mm	3–4 mm	2–3 mm	2–6 cm	purple-black	6 mm	densely hairy
R. nevadense	reddish	5 mm	4–5 mm	3–4 mm	3–7 cm	blue-black	8 mm	lacking or sparse
R. viscosissimum	white	10 mm	2 mm	4 mm	3–8 cm	black	8–10 mm	glandular

Fig. 18. *Carpenteria californica*

about 1–3 m tall, with simple opposite leaves, 5–8 cm long. Flowers are large, with white petals 2–3 cm long. The fruit is a capsule 10–12 mm long.

Relatively few shrubs are endemic in the Sierra Nevada. This is one of the most restricted but most spectacular ones, occurring in the Foothills Woodland and Yellow Pine Forest in Fresno County between the San Joaquin and Kings rivers. Carpenteria can be seen in May through July along the highway from Fresno to Shaver and Huntington lakes.

Jamesia or Cliff Bush (*Jamesia americana*). A much-branched shrub, 2–10 dm tall; ovate to rounded leaves, sharply toothed and densely hairy, especially on the lower surface. The rose-pink flowers are in dense clusters at the ends of twigs. The fruit is a small capsule. Jamesia occurs most commonly at elevations of 9000 to 12,000 feet, but occasionally as low as 7200 feet, in rocky areas from Tulare County to Fresno County.

[40]

Fig. 19. *Philadelphus lewisii*

Mock Orange (*Philadelphus lewisii*) (fig. 19). A rather diffuse, deciduous shrub to about 3 m tall. Leaves are opposite, egg-shaped, 3–8 cm long. Flowers are copious and conspicuous with white petals to 1–1.5 cm long. Fruits are much less conspicuous, being small capsules about 5 mm long. Mock Orange occurs mainly in the Yellow Pine forest below about 4500 feet from Tulare County north. *Philadelphus, Jamesia* and *Carpenteria* are sometimes placed in the related family, Hydrangeaceae, although obviously related to the Saxifragaceae.

Currant, gooseberry *(Ribes)*

Ribes is a large genus with some 31 species in California. Fifteen of these occur in the Sierra Nevada. Some are evergreen, others are deciduous. All are small shrubs, rarely over 2 m tall, with rather stiff branches. The berries are edible, but some are rather sour. The

Fig. 20. *Ribes roezlii*

spiny plants are the gooseberries; the plants lacking spines are the currants. All have simple, alternate, 3–5-lobed leaves. The flowers are small and white to pink, or lavender, rarely yellow.

Ribes species are the alternate host for the White Pine Blister Rust, a fungus that does no harm to the *Ribes*, but does considerable harm to the commercially important Western White Pine. During the 1940's an attempt was made to control White Pine Blister Rust by having crews go through the forest and root out the *Ribes* plants; this probably affected to some extent the original distribution and abundance of some of the *Ribes* species.

Table 7

Bitter Gooseberry (*Ribes amarum*). A rather erect shrub, 1–2 m tall. Occurs in Chaparral, Foothills Woodland, and Yellow Pine Forest, below about 5500 feet, on the western slopes of the Sierra Nevada north to El Dorado County.

Straggly Gooseberry (*Ribes divaricatum*). Height 1–3 m tall. Occurs in moist shaded habitats in forests from 3500 to 11,000 feet in the montane forests from Tulare County north.

Alpine Gooseberry (*Ribes lasianthum*) (pl. 2, *b*). A low, spreading shrub to 1 m high. Occurs in open rocky places, from Tulare County to Nevada County, generally at elevations between 7000 and 10,000 feet.

Canyon Gooseberry (*Ribes menziesii*). Height is 1–2 m. Occurs in Chaparral in the Sierra Nevada foothills below about 4000 feet in Fresno and Tulare counties.

Alpine Prickly Currant (*Ribes montigenum*). A straggly shrub, rarely more than 6 cm high. It occurs on dry rocky habitats between about 7000 and 12,000 feet throughout the Sierra Nevada.

Oak Gooseberry (*Ribes quercetorum*). A rounded shrub, 0.5–1.5 m tall, occurring at elevations below 4000 feet from Kern to Tuolumne conuties in the Foothills Woodland.

Sierra Gooseberry (*Ribes roezlii*) (fig. 20; pl. 2, *c*). Height 0.5–1 m, stiffly branched. It is widespread in the Sierra Nevada in the Yellow Pine and Red Fir forests from about 3500 to 8500 feet, usually in dry slopes.

Sequoia Gooseberry (*Ribes tularensis*). A low-branching shrub, usually under about 5000 and 6000 feet in the Yellow Pine and Red Fir Forests.

Plateau Gooseberry (*Ribes velutinum*). A very rigidly branched shrub, 0.5–2 m tall. It is found on the east-

Fig. 21. *Ribes cereum*

ern slopes of the Sierra Nevada from Inyo County north between 2500 and 8500 feet on dry slopes.

Table 8

Golden Currant (*Ribes aureum*) (pl. 2, *d*). Distinctive because of its yellowish flowers. It is known from Fresno and Inyo counties north in a number of plant communities between about 3000 and 8000 feet.

White Squaw (*Ribes cereum*) (fig. 21). Occurs in dry, rocky and exposed places, from about 5000 to nearly 13,000 feet, throughout the Sierra Nevada.

Pink Squaw Currant (*Ribes inebrians*). Closely related to the preceeding, differing from it in having a glabrous style and very narrow bracts among flower clusters. It is rare in the Sierra Nevada, generally occurring at high altitudes in exposed places on the eastern side of the mountain range.

Chaparral Currant (*Ribes malvaceum*) (pl. 2, *e*). Rarely occurs above about 2500 feet on the western foothills of the Sierra Nevada of El Dorado County, in Chaparral and Foothills Woodland.

Sierra Nevada Currant (*Ribes nevadense*). Quite common from about 3000 feet to 10,000 feet, generally along streams and in other moist places.

Sticky Currant (*Ribes viscosissimum*). As the name implies, this shrub is very glandular, especially on the younger leaves. It is found in shaded areas between 5000 to 9500 feet from Tulare to El Dorado counties.

ROSE FAMILY (ROSACEAE)

Shrubs of the rose family in the Sierra Nevada comprise a very large and diverse group of species. Eighteen genera occur in the Sierra Nevada. The flowers of all members of the family are characterized by having a floral cup surrounding the ovary, to which the sepals, petals, and stamens are attached. In some cases petals may be lacking; and the number of parts of the ovary can vary from one to many. Leaves are generally very useful in determining the different species, as are fruit types, which are exceedingly variable, ranging from fleshy berries, to applelike fruits to dry indehiscent pods. In a number of species the fruits are edible, and have been in the past an important food source for Indians and settlers. They are food for many birds as well.

Table 9 gives some characteristics of the different genera in the rose family.

Chamise or Greasewood (*Adenostoma fasciculatum*) (fig. 22). One of the most common shrubs in California. It is usually found in Chaparral, and in some areas forms almost pure stands for hundreds of acres. Leaves are needlelike, grouped in clusters or fascicles. It has a well-developed basal burl, and after the periodic chaparral fires numerous shoots arise from these. The white flowers are small but are aggregated into elon-

TABLE 9. Characteristics of Genera in the Rose Family

	Plant habit and height	Plants spiny or not	Leaf duration	Kind of leaf	Leaf or leaflet shape	Leaf or leaflet texture	Leaf or leaflet* length	Leaf or leaflet margin	Petal color	Fruit type
Adenostoma	erect, 0.5–3.5 m	no	evergreen	simple	linear	leathery	4–10 mm	entire	white	dry akene
Amelanchier	erect, 1–3 m	no	deciduous	simple	ovate-elliptic	thin	2–4 cm	toothed	white	fleshy, applelike
Cercocarpus	erect, 2–9 m	no	evergreen	simple	elliptic, lanceolate	leathery	1–3 cm	entire or toothed	petals 0	dry akene with a long tail
Chamaebatia	low, spreading, 2–6 dm	no	evergreen	much divided	—	thick	2–10 cm	toothed	white	dry akene
Chamaebatiaria	spreading, 0.6–2 m	no	evergreen	much divided	—	thick	2–4 cm	toothed	white	dry, pod-like
Coleogyne	erect, 0.3–2 m	yes	deciduous	simple	linear to club-shaped	thick	5–15 mm	entire	petals 0	dry akene
Holodiscus	erect, 1–2 m	no	evergreen	simple	ovate-elliptic	thin	1–3 cm	toothed	white to cream	dry akene
Oemleria	erect, 1–5 m	no	deciduous	simple	oblong	thin	5–10 cm	entire	white	fleshy berry
Petrophytum	mat-forming, 3–8 dm	no	evergreen	simple	spatulate	thick	5–12 mm	entire	white	dry, podlike

Genus	Habit	Armed	Leaf persistence	Leaf composition	Leaf shape	Leaf texture	Leaf size	Leaf margin	Flower color	Fruit
Photinia	erect, 2–10 m	no	evergreen	simple	oblong	leathery	5–10 cm	toothed	white	fleshy, applelike
Physocarpus	erect, 1–2.5 m	no	deciduous	simple	round-ovate	thin	3–7 cm	toothed	white	dry, podlike
Potentilla	spreading, 2–12 dm	no	evergreen	3–7 leaflets	narrowly lanceolate	thick	0.5–2 cm*	entire	yellow	dry akene
Prunus	erect, 1–6 m	some-times	deciduous	simple	ovate	thin	1–8 cm	toothed	white to rose	fleshy, cherry-like
Purshia	erect, 1–3 m	no	evergreen	simple	wedge-shapes	leathery	0.5–3 cm	3-toothed at apex	cream-yellow	dry akene
Rosa	erect to trailing, 0.5–3 m	yes	deciduous	3–7 leaflets	ovate to elliptic	thin	1–4 cm*	toothed	white to pink	fleshy, "hips"
Rubus	erect orvinelike, 1–2 m	yes, no	evergreen or deciduous	simple or with 3–5 leaflets	round to ovate	thin	5–15 cm* and leaf	toothed	white	fleshy, rasp-berrylike
Sorbus	erect, 1–4 m	no	deciduous	11–13 leaflets	lanceolate	thin	3–6 cm*	toothed	white	fleshy, applelike
Spiraea	erect, 0.2–2 m	no	deciduous	simple	oblong-elliptic	thin	5–9 cm	toothed	white or rose	dry, podlike

Fig. 22. *Adenostoma fasciculatum*

gate clusters 4–12 cm long, hence a stand of chamise in flower is whitish in appearance.

Service Berry (*Amelanchier alnifolia*) (fig. 23). Has an edible fruit, purplish in color, which resembles a small apple about 5 mm in diameter. Flower parts are situated above the fruit. Leaves are conspicuously veined and have sharp teeth in the distal half. Service Berry is a fairly common species in the Sierra Nevada from Kern County north, in the Red Fir and Lodgepole Pine Forests.

Mountain Mahogany *(Cercocarpus)*
Two species of mountain mahoganies occur in the Sierra Nevada, both of which are shrubs or small rounded trees, 2–9 m tall. The leaves are small, thick, and ever-green. The distinctive feature of this genus is the long, feathery, persistent styles which grow to be 4–9 cm long in fruit, giving the whole shrub a silvery appearance.

Fig. 23. *Amelanchier alnifolia*

California Mountain Mahogany (*Cercocarpus betuloides*). Has toothed leaves with flat margins. It occurs in Chaparral and Foothills Woodland below about 5500 feet along the western slopes of the Sierra Nevada.

Curl-leaved Mountain Mahogany (*Cercocarpus ledifolius*) (fig. 24). Easily distinguished from the preceding species as it has entire leaves with conspicuously recurved margins. It occurs mainly on the eastern slopes of the Sierra Nevada in Sagebrush Scrub, Pinyon-Juniper Woodlands, and Subalpine Forests, from about 4000 to 10,000 feet.

Mountain Misery or Ket-ket-dizze (*Chamaebatia foliolosa*) (fig. 25). Easily recognized by its finely divided leaves and pungent odor. The foliage and young twigs are sticky to the touch. This low shrub often forms extensive dense stands in the Yellow Pine Forest,

Fig. 24. *Cercocarpus ledifolius*

where it looks much like a carpet. It occurs from Tulare County north, between about 2500 and 6500 feet, in the Yellow Pine Forest and Red Fir Forest on the western slopes of the Sierra Nevada.

Mountain Misery, in addition to its common name of Ket-ket-dizze, has acquired others, among them Bear Mat, Tarweed, Bear Clover, Running Oak, and Tobacco Plant. Since we do not have any sort of standardization of common names yet, this is a good example of why one must use botanical names when communicating about plants.

Fern Bush (*Chamaebatiaria millefolium*) (pl. 2, *f*). Has very finely divided leaves that resemble those of milfoil. The foliage is aromatic and has scalelike pubescence. Fern Bush looks much like Mountain Misery, but it grows only on the eastern slopes of the Sierra Nevada between about 3500 and 10,000 feet on dry

Fig. 25. *Chamaebatia foliosa*

rocky slopes in Sagebrush Scrub and Pinon-Juniper Woodland.

Blackbush (*Coleogyne ramosissima*). Distinctive among the Sierra Nevada shrubs because it is the only one with opposite leaves in the Rosaceae. It is intricately branched; the small leaves are in fascicles; branches are spinescent in age. Flowers lack petals, but the four sepals and the floral cup are yellowish to brownish. Blackbush is mainly a desert shrub, but occurs occasionally on the lower slopes of the eastern Sierra Nevada, below 5000 feet, from Inyo County south.

Cream Bush or Mountain Spray (*Holodiscus discolor*) (fig. 26). Occurs throughout much of the Sierra Nevada from about 4000 to 11,000 feet. It is a variable species, and a number of names have been given to what is here called *Holodiscus discolor*. The flowers are small but arranged in pyramidlike clusters, so that a

Fig. 26. *Holodiscus discolor*

shrub in flower is very conspicuous. Leaves are usually toothed in the distal half.

Oso Berry (*Oemleria cerasiformis*) (fig. 27). Has an elongate fruit which is berrylike with a bitter taste. It is often confused with *Prunus*, but, unlike that genus, it has its male and female flowers on different plants. Oso Berry occurs in Chaparral and in the Yellow Pine Forest, usually in moister sites in canyons and on slopes, from Tulare County north, usually below about 5500 feet in elevation.

Rock Spiraea (*Petrophytum casespitosum*). Closely related to *Spiraea* and sometimes considered to be a member of that genus. It differs mainly in being evergreen and prostrate. It occurs occasionally in rocky limestone areas in Fresno and Tulare counties.

Toyon or Christmas Berry (*Photinia arbutifolia*) (fig. 28). A very handsome shrub regularly found in

[52]

Fig. 27. *Oemleria cerasiformis*

cultivation. The small white flowers are borne in clusters at the ends of young twigs. Most produce a fruit like a small apple, which at maturity is bright red and is eaten by birds. Toyon occurs in Chaparral and Foothills Woodland from Tulare County north from about 500 to 4000 feet in eleveation. In some botanical works, this species is placed in the genus *Heteromeles*.

Pacific Ninebark (*Physocarpus capitatus*) (fig. 29). Has 3–5 lobed leaves, about as broad as they are long. Flowers are small and are aggregated into dense, head-like clusters. It occurs on moist banks and north-facing slopes from Tulare County north, usually below 4500 feet, in several plant communities.

Bush Cinquefoil (*Potentilla fruticosa*) (fig. 30). One of the few shrubby members of this genus. Leaves have 3–7 small leaflets whose margins are recurved; the whole leaf is covered with silky hairs. Petals are yellow, making this a very conspicuous plant of high-

TABLE 10. Species in the Genus *Prunus*

	Height	Habit	Leaf shape and length	Petal color and length	Inflorescences
P. andersonii	1–2 m	spinescent	lanceolate, 1–2 cm	rose, 5–6 mm	solitary flowers
P. demissa	1–5 m	not spiny	oblong, 3–8 cm	white, 5–6 mm	elongate clusters, with 12 or more flowers
P. emarginata	1–6 m	not spiny	oblong-elliptic, 2–5 cm	white, 5–7 mm	flat-topped clusters with 3–10 flowers
P. subcordata	1–3 m	spinescent	ovate-roundish, 2–5 cm	white, 4–6 mm	clusters of 3–4 flowers

TABLE 11. Species in the Genus *Rosa*

	Habit and height	Number of leaflets	Petal length	Prickles	Flower tube	Fruit diameter
R. californica	erect, 1–3 m	5–7	10–25 mm	usually recurved, stout	lacking bristles	10–15 mm
R. gymnocarpa	erect, about 1 m	5–7	8–12 mm	slender, straight	lacking bristles	5–10 mm
R. pinetorum	erect, 0.5–1 m	5	10–14 mm	slender, straight	lacking bristles	6–12 mm
R. spithamea	low, creeping, 1–3 dm	3–7	12–20 mm	straight	with glandular bristles	7–8 mm

Fig. 28. *Photinia arbutifolia*

er altitudes. It occurs from Tulare County north, in the plant communities from Lodgepole Pine Forest to Alpine Fell Fields, between 7000 and 12,000 feet. It often grows among granitic boulders.

Stone Fruits *(Prunus)*

Four species of *Prunus* occur in the Sierra Nevada. All of them are shrubs; in two cases they can be small trees. All four are deciduous. The cherrylike fruits are edible, but often very bitter. The leaves are usually at least slightly toothed and often bear conspicuous glands on the leaf-stalk or petiole. The characters given in Table 10 will help to distinguish the species.

Desert Peach (*Prunus andersonii*) (pl. 3, *a*). A spinescent shrub with small leaves clustered together in fascicles. It occurs along the eastern side of the Sierra Nevada from Kern County north in dry areas in Sagebrush Scrub and Yellow Pine Forests.

Fig. 29. *Physocarpus capitatus*

Western Choke Cherry (*Prunus demissa*) (fig. 31). Easily recognized by its elongate or racemose, inflorescences. It is occasional in the Sierra Nevada below about 8000 feet in moist areas.

Bitter Cherry (*Prunus emarginata*). Can be recognized by its flat-topped clusters of white flowers. It often forms extensive thickets and is occasionally a small tree. It occurs in a variety of habitats about 9000 feet in the Sierra Nevada.

Sierra Plum (*Prunus subcordata*) (pl. 3, *b*). Has spinescent twigs. It is a variable species and occurs from Kern County north, generally below 6000 feet, in several plant communities.

Antelope Bush or Bitterbrush (*Pursia tridentata*) (fig. 32). Easily recognized by its short, thick, apically trilobed leaves and cream-yellow flowers. It is an important browse plant, as one of the common names indicates. It occurs mainly on the eastern slopes of the

Fig. 30. *Potentilla fruticosa*

Sierra Nevada from 3000 to about 10,000 feet from Tulare County north, usually in dry habitats.

Rose *(Rosa)*

Almost everyone can recognize a rose. The four species in the Sierra Nevada have white to pink or rose petals. The petals are five in number and thus differ from most ornamental roses, in which the petals are very numerous. The leaves have 3–7 leaflets, and these are usually toothed. The so-called fruits of the rose, the hip, is a floral tube which surrounds several ovaries, each with one seed. Hips are edible, and some people make a tea from the ripe ones. Natural hybrids occur between different species of roses. Table 11 should enable one to distinguish among most of the species in the Sierra Nevada.

California Rose *(Rosa californica)* (fig. 33). An exceedingly variable species. The recurved prickles are prominent; flowers occur in flat-topped clusters. Calif-

Fig. 31. *Prunus demissa*

ornia Rose occurs frequently in moist habitats near streams, below 6000 feet, in several plant communities on the western slopes of the Sierra Nevada.

Wood Rose (*Rosa gymnocarpa*). Readily distinguished because the sepals drop off before the seeds are mature; in the other species, the sepals are persistent. Flowers are usually solitary. Wood Rose occurs from Fresno County north in shaded woods below 6000 feet in several plant communities.

Pine Rose (*Rosa pinetorum*). Usually has only five leaflets and is not more than 0.5 m high. Flowers are usually solitary. It occurs from Tulare County north in wooded areas between about 2000 and 6000 feet.

Ground Rose (*Rosa spithamea*). The lowest of the Sierra Nevada roses, rarely over 2 dm high, and creeping. The flowers are solitary, or a few in clusters. It oc-

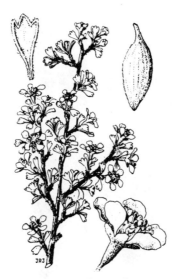

Fig. 32. *Purshia tridentata*

curs in open woods below 5000 feet from Tulare County north.

Raspberry, Blackberry, and Thimbleberry (*Rubus*). *Rubus* is a large genus of usually spiny, vinelike plants. In many parts of the world, the eastern United States and Europe for instance, they are considered to be a difficult group taxonomically, but in the Sierra Nevada, fortunately, they are easy to tell apart. The fruit is the typical raspberry or blackberry, with which every one is familiar, and they are all edible. Table 12 should allow one to distinguish between the three species which occur in the Sierra Nevada.

Western Raspberry (*Rubus leucodermis*) (fig. 34). Easily recognized by its leaves, which are whitish on the under surface due to a dense layer of hairs. It occurs from Tulare County north on moist slopes and in canyons in wooded areas, usually below about 6500 feet in elevation.

Fig. 33. *Rosa californica*

Thimbleberry (*Rubus parviflorus*) (pl. 3, *c*). Lacks prickles and has simple leaves which are round in outline and 5-lobed. It occurs throughout the Sierra Nevada below about 8000 feet in open woods.

California Blackberry (*Rubus ursinus*) (pl. 3, *d*). Vinelike, often forming dense, almost impenetrable thickets. *Rubus vitifolius* is here included within *Rubus ursinus*. It occurs commonly below about 4000 feet in a variety of habitats in the Sierra Nevada, from open to shaded areas.

Himalaya Berry (*Rubus procerus*). Naturalized at lower elevations in the Sierra Nevada. It has 3–5 leaflets. Its flowers are 2–2.5 cm across, while those of *Rubus leucodermis* are under 1 cm across.

Mountain Ash or Rowan (*Sorbus scopulina*). A shrub with several stems. Leaves have 11–13 leaflets;

Fig. 34. *Rubus leucodermis*

flowers are about 1 cm across with white petals, borne in clusters; the fruit is orange to scarlet and applelike, 8–10 mm in diameter. Mountain Ash occurs occasionally from Tulare County north in moist places and along waterways in coniferous forests from about 5000 to 9000 feet.

Spiraea *(Spiraea)*

Two species of spiraeas occur in the Sierra Nevada, both small shrubs with numerous branches. The flowers are small and arranged in dense clusters; the fruit consists of five separate, 1-chambered pods.

Mountain Spiraea (*Spiraea densiflora*) (pl. 3, *e*). Ranges in height from 3–9 dm. Flowers are pink, in flat-topped clusters. It occurs in rather rocky, forested areas between 5500 and 11,000 feet from Tulare County north.

Douglas' Spiraea (*Spiraea douglasii*). A larger shrub than Mountain Spiraea, 1–2 m tall. It has pink

[61]

TABLE 12. Species in the Genus *Rubus*

	Prickles	Number of leaflets	Lower surface of leaflets	Habit and height	Fruit color
R. leucodermis	present	3–5	white	vinelike, stems arched, to 2 m	purple-black
R. parviflorus	absent	1	green	erect, 1–2 m	red to scarlet
R. ursinus	present	3	green	vinelike, scrambling, to 2 m	black

flowers arranged in elongate clusters. Douglas' Spiraea occurs below 6000 feet in the extreme northern part of the Sierra Nevada.

Pea Family (Fabaceae)

The pea family is represented in the Sierra Nevada by many kinds of herbaceous plants and only a few shrubs. Generally, the leaves have several leaflets (*Cercis* being an exception); the flowers are irregular; and the fruit is usually like a pea pod. Of the four shrubby genera of legumes, *Cercis* can be distinguished from the others by its simple leaves. The other genera have compound leaves: *Lupinus* has 7–10 leaflets, while *Lotus* and *Pickeringia* usually have 3. *Lotus* has yellow flowers and is a low, nonspiny shrub; *Pickeringia* has purplish flowers and is a tall, spinescent shrub.

Red Bud (*Cercis occidentalis*) (fig. 35, pl. 3, f). A large deciduous shrub or rarely a small tree, with rounded, glossy leaves 3–9 cm long with 7–9 veins from the base of the leaf blade. Flowers have five reddish-purple petals, 8–12 mm long; fruits are flattened, 4–9 cm long at maturity.

This is a conspicuous shrub in the Sierra Nevada foothills from Tulare County north. The numerous fruits

[62]

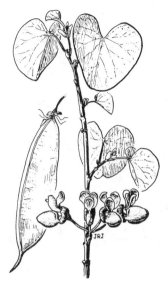

Fig. 35. *Cercis occidentalis*

remain on the shrubs for a long time and are a reddish color, thus giving the shrubs a decorated appearance.

Deer Weed (*Lotus scorparius*). A shrubby plant, usually less than 1.2 m tall, with many willowy branches from near the base of the plant. Leaves are small and rather inconspicuous. Yellow flowers are borne along the stems in clusters; the petals are 7–10 mm long and often turn reddish with age. Fruits are about as long as the petals and do not open to release the seeds. Deer Weed is common in poor soil in Chaparral, particularly after fires, generally below about 5000 feet on the western slopes of the Sierra Nevada.

Silver Lupine (*Lupinus albifrons*) (pl. 4, *a*). A low, much-branched shrubby plant to about 1.5 m tall. The leaves and branches are copiously covered with silky hairs, giving the plant a silvery appearance. The leaflets are 1–3 cm long. Flowers are in elongated clusters,

Fig. 36. *Ptelea crenulata*

generally blue to reddish-purple, 10–14 mm long. The fruit is a typical "pea pod," 3–5 cm long.

Most lupines are annual or perennial herbs. Among the perennial herbs a number have somewhat woody habits, but only one in the Sierra Nevada meets the criteria for a shrubby plant. Silver Lupine occurs in rocky places on dry hillsides in the open, from Kern County north below about 5000 feet, in several plant communities.

Chaparral Pea (*Pickeringia montana*) (pl. 4, *b*) A very spiny evergreen shrub to 2 m tall. Leaves consist of 3 small leaflets, each 4–12 mm long; flowers are purplish, 1.5–2 cm long, eventually producing a fruit 3–5 cm long. Chaparral Pea is among the easiest of the woody chaparral plants to learn, as it is the only one with large purplish pealike flowers. It occurs in the Sierra Nevada foothills in Chaparral from Mariposa County north.

[64]

a. *Tsuga mertensiana*

b. *Salix angelorum*

c. *S. exigua*

d. *S. lasiolepis*

e. *Quercus durata*

f. *Eurotia lanata*

Plate 1

c. *R. roezlii* (see fig. 20)

f. *Chamaebatiaria millefolium*

b. *Ribes lasianthum*

e. *R. malvaceum*

a. *Clematis lasiantha* (see fig. 13)

d. *R. aureum*

Plate 2

a. *Prunus andersonii*

b. *P. subcordata*

c. *Rubus parviflorus*

d. *R. ursinus*

e. *Spiraea densiflora*

f. *Cercis occidentalis* (see fig. 35)

Plate 3

a. *Lupinus albifrons*

b. *Pickeringia montana*

c. *Toxicodendron diversilobum*
(see fig. 37)

d. *Acer circinatum*

e. *Ceanothus fresnensis*

f. *C. velutinus*

Plate 4

a. *Opuntia basilaris*

b. *Kalmia polifolia*

c. *A. myrtifolia*

d. *A. nevadensis*

e. *A. viscida*

Plate 5

f. *Cassiope mertensiana*

a. *Gaultheria ovatifolia*

b. *Arctostaphylos mariposa*

c. *Leucothe davisiae*

d. *Phyllodoce breweri*

e. *Fraxinus dipetala*

f. *Leptodactylon pungens*

Plate 6

a. *Mimulus longiflorus*

b. *Penstemon breviflorus*

c. *P. bridgesii*

d. *P. davidsonii*

e. *P. newberryi*

f. *Lonicera interrupta*

Plate 7

a. *Sambucus melanocarpa*

b. *Symphoricarpos rivularis*

c. *Artemisia tridentata*

d. *Baccharis viminea*

e. *Haplopappus bloomeri*

f. *Tetradymia axillaris*

Plate 8

Fig. 37. *Toxicodendron diversilobum*

RUE FAMILY (RUTACEAE)

Hop Tree (*Ptelea crenulata*) (fig. 36). A member of the family to which all our citrus fruits belong, though the Hop Tree does not produce any edible products for man. It is a shrub or small tree, usually under 3 m tall, with alternate leaves composed of three leaflets. Flowers are small, but in dense clusters; fruits are 1–2 cm in diameter, bladdery, and with wings. Hop Trees occur on the western slopes of the Sierra Nevada below about 2000 feet from Tulare County north in the Foothills Woodland and Yellow Pine Forest.

SUMAC FAMILY (ANACARDIACEAE)

Two members of this family occur as shrubs in the Sierra Nevada. Poison Oak is a plant to know, respect, admire, and avoid—unless you are fortunate enough to be immune to its powers to cause you to itch, swell, and be uncomfortable. Squaw Bush is harmless.

[65]

Fig. 38. *Euonymus occidentalis*

Poison Oak (*Toxicodendron diversilobum*, formerly known as *Rhus diversiloba*) (fig. 37). An exceedingly variable species, ranging from small willowy shrubs, to small trees, to rather strikingly large vines. Leaves are alternate, deciduous, and have three egg-shape leaflets, 2–7 cm long. The leaflets may be entire margined or lobed. Flowers are small, and the male and female flowers are on different parts of the plant. Poison Oak is common in the Sierra Nevada below about 5000 feet, but occurs occasionally at high altitudes, in some plant communities. It may be confused with *Clematis*, a vine, but Poison Oak has alternate leaves, while those of *Clematis* are opposite. Poison Oak is one of the plants that provides us with fall color. (pl. 4, *c*).

Squaw Bush or Skunk Bush (*Rhus trilobata*). A low, dense subshrub to 1.5 m tall, with branches that droop at the tips. Leaves are much like those of Poison Oak, but smaller, the leaflets being only 3 cm long. Flowers

Fig. 39. *Staphlea bolanderi*

contain both male and female elements and are usually
yellowish. This shrub is occasional in Chaparral and
Foothills Woodlands on the western slopes of the Sierra
Nevada.

STAFF TREE FAMILY (CELASTRACEAE)
Western Burning Bush (*Euonymus occidentalis*) (fig.
38). A shrub of shaded areas, with deciduous opposite
leaves, 3–9 cm long. It rarely exceeds 4 m in height.
The flowers are small, with brownish-purple petals, 3–
4 mm long. One of the most distinctive features of
Burning Bush is the aspect of the fruits, usually
arranged in 3's and looking much like the traditional
pawnbroker's symbol. It is rare in the Sierra Nevada,
being known only from Plumas County.

Mountain Lover (*Paxistima myrsinites*). A low shrub
or subshrub, rarely even as much as 1 m high, with
small evergreen leaves, 1–2.5 cm long. The flowers are

Fig. 40. *Aesculus californica*

quite small, with reddish-brown petals about 1 mm long. Mountain Lover is fairly abundant but inconspicuous in mixed evergreen, Yellow Pine, and Red Fir Forests from Mariposa County north. In Oregon the common name would probably be Oregon Boxwood.

BLADDER-NUT FAMILY (STAPHYLEACEAE)
California Bladder-nut (*Staphylea bolanderi*) (fig. 39). A very distinctive shrub, rarely a small tree, 2–6 m tall, with opposite leaves composed of three leaflets, each leaflet egg-shaped, 2.5–6 cm long. Flowers are conspicuous, about 1 cm long, with white petals. The 3-lobed bladdery fruit is even more conspicuous, being 2.5–5 cm long and bearing 3 hornlike projections. The fruits persist for a long time on the plants. California Bladder-nut is occasional in Foothills Woodland, Chaparral, and Yellow Pine Forest, below about 4500 feet and above about 1000 feet, from Tulare County north.

[68]

Fig. 41. *Ceanothus cuneatus*

MAPLE FAMILY (ACERACEAE)

Vine Maple (*Acer circinatum*) (pl. 4, *d*). A small, deciduous shrub, sometimes vinelike, occasionally a small tree. Leaves are opposite, 5–11-lobed, rounded, with the major veins arising from one point, and 5–12 cm across. Flowers are small, 4–6 mm long, in clusters, with white to greenish petals. As with all maples, the fruits are very distinctive, and consist of 2-winged capsules, 2–3 cm long. Vine Maples are very colorful in the fall and occur in partial shade along streams from Yuba County north, usually below 5000 feet.

Mountain Maple (*Acer glabrum*). Usually a tree, but occasionally shrublike. Leaves are small, 2–4 cm wide with 3 lobes. It occurs mostly above 5000 feet in the Sierra Nevada.

HORSE-CHESTNUT OR BUCKEYE FAMILY (HIPPOCASTANACEAE)

California Buckeye (*Aesculus californica*) (fig. 40). Usually considered a tree, but some individuals are

[69]

large shrubs, much branched from near the base. The bark is grayish-white and often covered with lichens. Leaves have 5–7 leaflets, all arising from the end of the leaf stalk; each leaflet is 5–15 cm long. Flowers are white to pink, about 1.5 cm long, in elongate spikelike clusters, 10–20 cm long, at the tips of the smaller branches, giving the whole plant the appearance of having many candles covering it. From each cluster of flowers, usually only one or two fruits develop. These are chestnutlike, 3–5 cm across.

California Buckeye reflects the seasons. The leaves appear in midwinter and are shed by the time it is hot and dry in the middle of the summer. In the fall the branches are bare except for the ornamentlike ripening fruits. It is usually found in the Foothills Woodland below about 4000 feet on the western slopes of the Sierra Nevada, often in rather dry sites.

BUCKTHORN FAMILY (RHAMNACEAE)

Two genera, *Ceanothus* and *Rhamnus*, of the buckthorn family are important elements of the shrub flora of the Sierra Nevada, particularly in Chaparral. The two can be distinguished quite easily. Some of the more important differences are given in Table 13.

TABLE 13. Comparison between the Genera *Ceanothus* and *Rhamnus*

	Leaf position	Leaf veination	Petals	Fruit
Ceanothus	alternate or opposite	1–3 main veins from the base	always present, hooklike, much narrowed toward the base	dry woody capsule
Rhamnus	alternate	1 main vein from the base	lacking or present	fleshy berry

TABLE 14. Species of *Ceanothus* with Opposite Leaves

	Habit and height	Horns on capsules	Flower color	Leaf length	Leaf margin
C. cuneatus	erect, 1–3.5 m	present, conspicuous	white	5–15 mm	teeth lacking
C. fresnensis	prostrate, 3 dm	lacking or short	blue	6–12 mm	toothed at the tip
C. pinetorum	prostrate to erect, 2–14 dm	present, conspicuous	white to blue	12–25 mm	coarse teeth present
C. prostratus	prostrate, creeping, 1.5 dm	present, conspicuous	blue	8–25 mm	3 prominent teeth at tip of leaf
C. vestitus	erect, 1–2 dm	lacking or short	white	10–20 mm	conspicuously toothed

TABLE 15. Species of *Ceanothus* with Alternate Leaves with One Main Vein from the Base

	Habit and height	Horns on capsule	Flower color	Leaf length	Leaf margin	Flower cluster length
C. diversifolius	trailing, 1–3 dm	small crests	white to blue	0.5–2 cm	toothed	1 cm
C. integerrimus	erect, 1–4 m	small crests	white, blue, or pink	2.5–7 cm	not toothed	4–15 cm
C. lemmonii	low, spreading, 3–8 dm	conspicuous horns	pale blue	0.8–3 cm	toothed	1–3 cm
C. palmeri	erect, 1–3.5 m	horned	white	1.5–3.5 cm	not toothed	7–12 cm
C. parvifolius	low, spreading, 6–12 dm	not, or slightly, crested	pale to deep blue	0.6–2 cm	not toothed	3–7 cm

TABLE 16. Species of *Ceanothus* with Alternate Leaves with Three Main Veins from the Base

	Habit and height	Horns on capsule	Flower color	Leaf length	Leaf margin	Flower cluster length
C. cordulatus	erect, 1–2 m	small crests	white	1–2 cm	usually not toothed	1.5–3 cm
C. integerrimus	erect, 1–4 m	small crests	usually white, but also blue or pink	2.5–7 cm	not toothed	4–15 cm
C. leucodermis	erect, 2–4 m	lacking	white to pale blue	1–2.5 cm	toothed or not	3–8 cm
C. parvifolius	low, spreading, 6–12 dm	not, or slightly crested	pale to deep blue	0.6–2 cm	not toothed	3–7 cm
C. tomentosus	erect, 1–3 m	lateral crests	white to blue	1–2.5 cm	finely toothed	2–5 cm
C. velutinus	erect, 1–2 m	lacking or minute	white	2.5–8 cm	finely toothed	5–10 cm

California Lilac (*Ceanothus*)

Ceanothus, with *Arctostaphylos* and *Salix*, is one of the most diverse groups of shrubs in the vegetation of the Sierra Nevada. There are nearly 50 species of *Ceanothus* in California alone. The attractive, fragrant floral clusters of *Ceanothus* are well known to native plant enthusiasts, and many species are cultivated very often as ground covers. Numerous hybrids occur. Many kinds of *Ceanothus* respond to periodic natural fires by producing many seedlings or by sprouting from underground parts after the fire. California Lilac is an important part of Chaparral.

For ease in telling them apart, the California Lilacs are divided into three groups, which are described in Tables 14–16. Note that in a few cases the same species may appear in more than one of the following groups.

I. *Ceanothus* species with opposite leaves. (Table 14).

Common Buck Brush (*Ceanothus cuneatus*) (fig. 41). A common shrub below about 6000 feet throughout the Sierra Nevada in Chaparral, Yellow Pine Forest, and Pinon-Juniper Woodland.

Fresno Ceanothus (*Ceanothus fresnensis*) (pl. 4, *e*). Endemic in the Sierra Nevada in the Yellow Pine Forest in dry areas between 3000 and 6000 feet from Fresno to Tuolumne counties.

Coville's Ceanothus (*Ceanothus pinetorum*). Occurs on dry slopes between 5000 and 9000 feet in Yellow Pine, Red Fir, and Lodgepole Pine Forests from Fresno and Mono counties south.

Squaw Carpet or Mahala Mats (*Ceanothus prostratus*) (fig. 42). Occurs in open flats in the Yellow Pine and Red Fir Forests between about 3000 and 7000 feet from Calaveras and Alpine counties north.

Fig. 42. *Ceanothus prostratus*

Mohave Ceanothus (*Ceanothus vestitus*). Occurs on dry slopes in Sagebrush Scrub on the eastern slopes of the Sierra Nevada from Mono County south, between 3500 and 7500 feet.

II. *Ceanothus* species with alternate leaves which have one main vein from the base (Table 15).

Pine Mat (*Ceanothus diversifolius*). Occurs occasionally in the Yellow Pine Forest from about 3000 to 6000 feet, on the western slopes of the Sierra Nevada from Kern County north.

Deer Brush (*Ceanothus integerrimus*) (fig. 43). A very distinctive mid-Sierran species with elongate spike s of flowers. Leaves sometimes with only one main vein from the base. It occurs mainly in the Yellow Pine Forest from about 2500 to 7000 feet.

[74]

Fig. 43. *Ceanothus integerrimus*

Lemmon's Ceanothus *(Ceanothus lemmonii)*. Ranges from Tuolumne County north in the Yellow Pine Forest and Foothills Woodland in open areas at elevations of 1000 to 3500 feet.

Palmer's Ceanothus *(Ceanothus palmeri)*. Occurs in the Sierra Nevada in Amador and El Dorado counties on dry slopes in Chaparral and Foothills Woodland between about 500 and 1500 feet elevation.

Small-leaved Ceanothus *(Ceanothus parvifolius)*. Leaves sometimes with only one main vein from the base. A plant of wooded areas between 4500 and 7500 feet from Tulare County north.

III. *Ceanothus* species with alternate leaves which have three main veins from the base (Table 16).

Snow Bush *(Ceanothus cordulatus)*. Common in the Sierra Nevada in open habitats in a number of plant communities, but mainly above the Yellow Pine Forest from about 4000 to 10,000 feet.

[75]

TABLE 17. Species in the Genus *Rhamnus*

	Leaf duration	Plant height	Leaf length	Petals	Flowers	Berry color	Berry diameter
R. alnifolia	deciduous	1–1.5 m	1–10 cm	lacking	unisexual	black	6–8 mm
R. californica	evergreen	1–4 m	3–8 cm	present	bisexual	red to dark red	10–12 mm
R. crocea	evergreen	1–4 m	1–4 cm	lacking	unisexual	red or black	5–6 mm
R. purshiana	deciduous	1–1.5 m	8–20 cm	present	bisexual	black	10 mm
R. rubra	deciduous	4–12 m	3–7 cm	present	bisexual	black	8–10 mm

Fig. 44. *Rhamnus californica*

Chaparral Whitethorn (*Ceanothus leucodermis*). Occurs from Eldorado County south, below about 6000 feet, in Chaparral and Foothills Woodland.

Wooly-leaved Ceanothus (*Ceanothus tomentosus*). Ranges in a restricted territory, Mariposa County to Placer County in the Sierra Nevada below about 5000 feet, in Chaparral and the Yellow Pine Forest.

Tobacco Bush (*Ceanothus velutinus*) (pl. 4, *f*). Occurs in wooded areas between 3000 and 10,000 feet from Tulare County north in several plant communities.

Buckthorn or Coffeeberry (*Rhamnus*)

The five species of Buckthorn or Coffee berry in the Sierra Nevada are shrubs or small trees with alternate leaves. The flowers are small and inconspicuous, and some species lack petals. The fruit is a fleshy 2–3-lobed berry, 5–12 mm in diameter. It is often difficult to ident-

[77]

Fig. 45. *Rhamnus crocea*

ify individual plants of this genus because hybrids occur. The set of characters in Table 17 should help identify most specimens of *Rhamnus*.

Alder-leaved Coffeeberry (*Rhamnus alnifolia*). Occurs in the Red Fir and Lodgepole Pine Forests from Placer County north between about 4500 and 7000 feet.

California Coffeeberry (*Rhamnus californica*) (fig. 44). Occurs on both slopes of the Sierra Nevada, usually between 1000 and 7500 feet in several plant communities.

Redberry (*Rhamnus crocea*) (fig. 45). Occurs below about 5000 feet in several plant communities throughout the foothills of the Sierra Nevada. It usually has the smallest leaves of any of the *Rhamnus* species.

Cascara Sagrada (*Rhamnus purshiana*). Often a small tree, as well as a shrub. It occurs from Placer

[78]

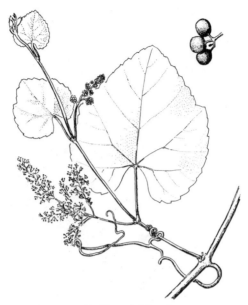

Fig. 46. *Vitis californica*

County north in the Yellow Pine Forest between 3000 and 5000 feet. The bark has been used extensively by both Indians and white men as a source of laxatives.

Sierra Coffeeberry (*Rhamnus rubra*). Occurs from Mariposa and Tuolumne counties north between 2000 and about 7000 feet in several plant communities.

GRAPE FAMILY (VITACEAE)

California Wild Grape (*Vitis californica*) (fig. 46). A woody vine, to 10 m tall, climbing by means of tendrils which are opposite the large, lobed leaves. Leaves have a typical "grapelike" shape, and are 7–16 cm long; flowers are small, about 1.5 mm long, in clusters; fruits (grapes) are 6–10 mm long, purplish, and without much pulp.

Cultivated grapes in the San Joaquin Valley of California produce leaves that are picked, canned, and sold for use in Greek recipes which ask for vine leaves. Leaves of the California Wild Grape will serve the

Fig. 47. *Fremontodendron californicum*

same purpose, for those who are willing to take the trouble to gather them.

California Wild Grape occurs below about 4000 feet mainly in the Foothills Woodland and Yellow Pine Forest from Kern County north. It is well represented along the Merced River below Yosemite National Park.

Mallow Family (Malvaceae)

Fremont's Globe Mallow (*Malacothamnus fremontii*). An erect shrub 1–2 m tall, with leaves 2.5–10 cm long having up to 7 shallow lobes. Young parts of the plant, including the leaves, are covered with dense mats of hairs, which impart a white color. The pinkish-rose flowers are in elongate clusters at the tips of new growth. As is characteristic in the Mallow Family, the stamens form a tube around the styles. This species is scattered throughout the Foothills Woodland and Chaparral from Tulare to Amador counties, usually below 3000 feet.

Fig. 48. *Shepherdia argentea*

CACAO FAMILY (STERCULIACEAE)

Two species of flannel bushes occur in the Sierra Nevada. Both are small shrubs with large conspicuous flowers, which lack petals. The sepals provide the color to the flowers. In both species the stamens are numerous and united into a tube around the style. The fruits are hairy, 5-parted capsules, 2–3.5 cm long at maturity. The whole plant is covered with a dense layer of star-shaped hairs. The genus is named in honor of the early western explorer and military man, John Charles Fremont.

California Fremontia (*Fremontodendron californicum*) (fig. 47). A much-branched erect shrub, 1.5–4 m tall with 3-lobed, rounded leaves to 3 cm long. The flowers are clear yellow, 3.5–6 cm across. California Fremontia occurs mainly on granitic soils on the western slopes of the Sierra Nevada between about 3000 to 6000 feet in Chaparral and Yellow Pine Forest.

Lloyd's Fremontia (*Fremontodendron decumbrens*). A low woody sprawling shrub, rarely over 1 m high. It

[81]

TABLE 18. Species in the Genus *Cornus*

	Leaf length (cm)	Leaf width (cm)	Petal like bracts	Bract length (cm)	Flowers and fruits
C. glabrata	3–5	1.5–2.5	absent	——	in loose clusters
C. nuttallii	6–12	3–7	present	4–6	in dense heads
C. sessilis	4–9	2–2.5	present	1.5	in heads
C. stolonifera	5–9	1.5–5	absent	——	in loose heads

has 5–7 deeply lobed leaves, 2–3 cm long; the flowers are a striking red-orange to red-yellow, 3–3.5 cm across. This is one of the rarest of Sierra Nevada shrubs, being known only from near Rescue in El Dorado County in oak-pine areas at about 1900 feet elevation. It was discovered in 1956, and described by Dr. Robert M. Lloyd in 1965.

Cactus Family (Cactaceae)

Beaver Tail Cactus (*Opuntia basilaris*) (pl. 5, *a*). This cactus barely fits the definition of a shrub, but it does form low bushy plants. Leaves are vestigial, hence to all purposes "lacking." The photosynthetic function of leaves in most plants is here assumed by the flattened, fleshy, jointed, spiny stems. Flowers are white to red, with numerous, very narrow petals. This cactus occurs on the eastern slopes of the Sierra Nevada in Mono and Inyo counties in rocky places in desertlike habitats.

Oleaster Family (Elaeagnaceae)

Buffalo Berry (*Shepherdia argentea*) (fig. 48). Occurs as rounded shrubs, occasionally small trees, 2–6 m tall. Leaves are opposite and silvery on both surfaces, due to small, silver overlapping scales. These scales also occur on young twigs, and are gradually shed as secondary growth occurs. Flowers are inconspicuous and are separated on male and female plants. The fruits are small, reddish, edible berries, 4–6 mm long.

[82]

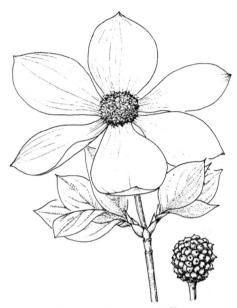

Fig. 49. *Cornus nuttallii*

Buffalo Berry occurs along the eastern slopes of the Sierra Nevada in Sagebrush Scrub and Pinon-Juniper Woodland between 3000 and 7000 feet.

A related introduced species, the Russian Olive (*Elaeagnus angustifolius*), is occasionally encountered as a shrub or small tree in the vicinity of old habitations or other marks of man's activities.

Dogwood Family (Cornaceae)

Four species of dogwoods occur in the Sierra Nevada. All lose their leaves in the fall. Once you can recognize one dogwood from the distinctive leaves, you will be able to recognize other members of the genus on sight. Table 18 will help distinguish the four species.

Smooth Dogwood (*Cornus glabrata*). A shrub or small tree up to 6 m tall. The inconspicuously veined leaves are usually oval in shape; the flower clusters produce white or bluish fleshy fruits with one hard seed. Smooth Dogwood grows along stream banks and in

Fig. 50. *Cornus stolonifera*

other moist sites below about 5000 feet in a number of plant communities.

Mountain Dogwood (*Cornus nuttallii*) (fig. 49). A very striking shrub or small tree. The flowers, subtended by 4–7 petal-like white to pinkish bracts, appear before the leaves; leaves are hairy on both surfaces; fruits at maturity are red and form a dense head. This dogwood grows below about 6000 feet, from Tulare County north in the Sierra mixed coniferous forests.

Black Fruit Dogwood (*Cornus sessilis*). A shrub or small tree, 1–4 m high. The lower surfaces of the leaves have scattered hairs; flowers are clustered and appear either before or with the emergence of the leaves; the fruit at maturity is shiny black. This dogwood is found along stream banks from about 1000 to 5000 feet from Calaveras County north.

American Dogwood (*Cornus stolonifera*) (fig. 50). A spreading shrub 2–5 m high. Leaves are conspicu-

[84]

Fig. 51. *Garrya fremontii*

ously veined, with up to 7 lateral veins on each side of the midvein; fruits are white at maturity. This shrub occurs below about 9000 feet in the montane coniferous forests in moist places.

SILK-TASSEL FAMILY (GARRYACEAE)

Congdon's Silk-tassel (*Garrya congdonii*). A shrub which may reach 2 m in height. The yellow-green leaves are 2.5–7 cm long, the margins are both entire and wavy, and the undersurfaces have a dense covering of long hairs. As in all kinds of silk-tassel, the male and female flowers are on different plants; the female tassels are usually shorter than the male. Hairs on the mature fruits give them a silky texture. This species occurs in the foothills of the Sierra Nevada from Mariposa County north below about 3000 feet in Chaparral and Foothills Woodland.

Ashy Silk-tassel (*Garrya flavescens*). A shrub 1.5–2.5 m tall. Leaves are grayish or ashlike in color, about 3–

TABLE 19. Genera in the Heath Family

Genus	Plant height	Leaf duration	Leaf position	Leaf length	Corolla color	Corolla shape	Corolla length	Fruit position above or below other flower parts
Arctostaphylos	0.1–4 m	evergreen	alternate	0.5–7 cm	white to pink	urn	4–8 mm	above
Cassiope	1–3 dm	evergreen	opposite	3–6 mm	white to pink	bell	5–6 mm	above
Gaultheria	2 dm	evergreen	alternate	1–2 cm	white	bell	5–7 mm	above
Kalmia	1–2 dm	evergreen	opposite	1–2 cm	rose-purple	saucer	8–12 mm	above
Ledum	0.5–1.5 m	evergreen	alternate	1.5–3 cm	white	petals separate	5–8 mm	above
Leucothoe	0.5–1.5 m	evergreen	alternate	3–6 cm	white	urn	6–7 mm	above
Phyllodoce	1–3 dm	evergreen	alternate	1 cm	rose-purple	bell	10 mm	above
Rhododendron	1–3 dm	deciduous	alternate	3–8 cm	pink	funnel	3.5–5 cm	above
Vaccinium	0.1–4 m	evergreen	alternate	1–3 cm	white to pink	bell	4–6 mm	below

6 cm long, with entire, flat margins. Fruits are densely hairy. This species is common on the more arid slopes of the southern Sierra Nevada from Tulare County to Fresno and Inyo counties.

Fremont's Silk-tassel (*Garrya fremontii*) (fig. 51). An erect shrub, 1–3 m tall, with yellowish green leaves, 1.5–8 cm long with flat, entire margins. The female tassels are about 10 cm long in fruit; male tassels are often over 20 cm long. Fremont's Silk-tassel occurs below about 7500 feet in the Sierra mixed coniferous forest, Yellow Pine Forest and in Chaparral.

HEATH FAMILY (ERICACEAE)

The members of the heath family range in size from low subshrubs to large trees. Some, like *Cassiope*, are low woody plants and barely fit the general description of a shrub, but are included because most works on shrubs do include them. The corollas are usually showy and are variously bell-shaped, urn-shaped, saucer-shaped, or lobed to near the base. The leaves are usually leathery and with entire margins.

Nine genera of the heath family occur in the Sierra Nevada. These can be distinguished by using Table 19.

Manzanita (*Arctostaphylos*)

The manzanitas comprise one of the largest groups of shrubs in California, there being probably about 50 described species. In some cases, botanists, lacking an understanding of current theory, have complicated our understanding of the manzanitas by describing local populations as species. Such very local species of manzanitas are properly described as endemics. Other species have very much wider ranges of distribution; *Arctostaphylos uva-ursi*, for example, is circumpolar in the northern hemisphere, though it happens to be very rare in the Sierra Nevada.

The bark of the manzanitas provides an identifying feature: on old stems it is usually a very distinctive reddish-brown. Some kinds of manzanita are charac-

TABLE 20. Species in the Genus *Arctostaphylos*

	Burl	Habit	Height	Leaf shape	Leaf length	Leaf base	Leaf color	Corolla length	Ovary
A. *manzanita*	lacking	erect	2–4 m	oblong to elliptic	2.5–4.5 cm	rounded	bright green	7–8 mm	glabrous to slightly hairy
A. *mariposa*	lacking	erect	1–4 m	round, ovate, to elliptic	2.5–4 cm	rounded	gray-green	6–7 mm	glandular, sticky
A. *mewukka*	present	erect	1–2.5 m	oblong to elliptic	2.5–4 cm	rounded to wedge-shaped	pale gray-green	6–7 mm	without hairs
A. *myrtifolia*	lacking	erect	3–8 dm	elliptic to narrowly ovate	0.5–1.5 cm	obtuse to acute	light green	4 mm	with stiff short hairs
A. *nevadensis*	lacking	low, prostrate	3–6 dm	lanceolate to elliptic	2–2.5 cm	wedge-shaped	light green	6–7 mm	without hairs
A. *nissenana*	lacking	erect	0.6–1.5 m	elliptic-oblong	1–2.5 cm	obtuse to truncate	pale green	4–5 mm	hairy
A. *patula*	present	erect	1–2 m	ovate to rounded	2.5–4 cm	rounded	light green	5–8 mm	without hairs
A. *uva-ursi*	lacking	prostrate	1–2 dm	ovate to obovate	1–2.5 cm	wedge-shaped	shiny, dark green	4–5 mm	without hairs
A. *viscida*	lacking	erect	1–4 m	ovate to elliptic	2.5–4 cm	truncate	whitish green	6–7 mm	without hairs to glandular
A. *truei*	lacking	erect	1–3 m	broadly ovate	4–7 cm	rounded to wedge-shaped	gray-green	7–8 mm	without hairs

terized by extensive underground burls from which new shoots arise after the periodic natural fires. Leaves are very thick, often quite leathery in texture, and are long persistent. The flowers occur in elongate clusters at the ends of young twigs; corollas are urn-shaped, white or tinged with pink. The fruit is fleshy and edible, although not palatable to all tastes. Seeds within the fruits are either separate, or united into a hard "stone."

Some manzanitas occur in relatively dry habitats, and increasingly there is an interest in using selected strains as horticultural plants. In some parts of California, the burls of manzanita were used as a source of material for pipes during World War II, when the supply of briar, also a member of the heath family, was cut off in the Mediterranean region.

Table 20 gives some of the characteristics by which manzanitas can be distinguished.

Parry Manzanita (*Arctostaphylos manzanita*). Occurs in Chaparral, Foothills Woodland, and Yellow Pine Forest between about 300 and 4000 feet from Mariposa County north, often forming extensive dense stands.

Mariposa Manzanita (*Arctostaphylos mariposa*) (pl. 6, *b*). Occurs in the Yellow Pine Forest from Kern County north to Amador County.

Indian Manzanita (*Arctostaphylos mewukka*). Ranges from about 2500 to 6000 feet in chaparral and the Yellow Pine Forest from Tulare County north.

Ione Manzanita (*Arctostaphylos myrtifolia*) (pl. 5, *c*). Restricted to rocky ridges and slopes in Amador and Calaveras counties between about 300 and 800 feet in Chaparral. It is very common in the vicinity of Ione, Amador County, hence the common name.

Pinemat Manzanita (*Arctostaphylos nevadensis*) (pl. 5, *d*). The most common prostrate or low sprawling manzanita in the Sierra Nevada, occurring between

Fig. 52. *Arctostaphylos patula*

5000 and 10,000 feet from Kern County north. It is more common toward the upper part of its altitudinal range.

Eldorado Manzanita (*Arctostaphylos nisseana*). One of the more localized manzanitas, being known from only a few places near Placerville in El Dorado County, in dry places in Chaparral and Foothills Woodland between about 1500 and 3500 feet.

Greenleaf Manzanita (*Arctostaphylos patula*) (fig. 52). Widespread from Kern County north in open places between 2000 and 11,000 feet in a number of plant communities.

Bear Berry (*Arctostaphylos uva-ursi*). Known to occur only in the Convict Lake Basin on the eastern slopes of the Sierra Nevada in Mono County, on decomposed marble at about 8000 to 10,000 feet.

White Leaf Manzanita (*Arctostaphylos viscida*) (pl. 5, *e*). A common species along the western slopes of the Sierra Nevada in Chaparral, Foothills Woodland, and Yellow Pine Forest, from about 500 to 5000 feet.

True Manzanita (*Arctostaphylos truei*). Found locally in Plumas and Yuba counties between 1500 and 3500 feet in chaparral.

White or Western Mountain Heather (*Cassiope mertensiana*) (pl. 5, *f*). Another low, creeping member of the heath family, rarely over 3 dm high. Leaves are small, 3–6 mm long, thick, with a ridge on the back, and arranged in four rows on the stems. Flowers are white to pinkish, bell-shaped, 5–6 mm long. Western Mountain Heather is most common in crevices in rocky areas, often forming extensive mats. It is found from Fresno County north, usually above 10,000 feet.

Alpine Wintergreen (*Gaultheria humifusa*). A matted, low woody plant, rarely over 2 dm high. Leaves are 1–2 cm long, egg-shaped, and leathery. Flowers are bell-shaped, white, 5–7 mm long, and occur singly in the leaf axils. Alpine Wintergreen is probably often overlooked because of its low stature. It occurs from 8000 to 11,000 feet in elevation, from Tulare County to Eldorado County.

Oregon Spicy Wintergreen (*Gaulthera ovatifolia*) (pl. 6, *a*). Related to Alpine Wintergreen. It occurs in a few localities in the Sierra Nevada in Sierra and El Dorado Counties in wet places, between 3000 and 5000 feet. Oregon Spicy Wintergreen has hairs on the calyx; Alpine Wintergreen does not.

American Laurel (*Kalmia polifolia*) (pl. 5, *b*). A low subshrub, rarely more than 1–2 dm high, with thick, opposite leaves, 1–2 cm long. The margins of the leaves are rolled under. Flowers are rose-purple with saucer-

Fig. 53. *Ledum glandulosum*

shaped corollas, 8–12 mm across. It is most common in wet meadows between about 7000 and 12,000 feet.

Labrador Tea (*Ledum glandulosum*) (fig. 53). A low shrub 0.5–1.5 m tall with thick, leathery leaves, 1.5–3 cm long, usually yellow-green beneath. Flower petals are white and separate. The leaves have been used to make a tea, hence the name. Labrador Tea occurs in boggy places and along streams from Tulare County County north, generally above 6000 feet.

Sierra Laurel (*Leucothoe davisiae*) (pl. 6, *c*). An erect shrub, 0.5–1.5 m tall, with leaves 3–6 cm long. The white flowers are in elongate inflorescences; the corollas are urn-shaped, 6–7 mm long. Like many members of the heath family, it occurs in boggy areas, from Fresno County north, in the Red Fir and Lodgepole Pine Forests, generally from 5000 to 8500 feet.

Red or Purple Mountain Heather (*Phyllodoce brew-*

Fig. 54. *Rhododendron occidentale*

eri) (pl. 6, *d*). A distinctive low heathlike plant, 1–3 dm high, with slender, needlelike leaves, usually about 1 cm long. Flowers are borne in small clusters at the tips of the branches; the corollas are rose-purple, widely bell shaped, and about 1 cm long. Purple Mountain Heather occurs in rocky areas, often in crevices between granitic boulders, from Tulare County north, usually above about 10,000 feet.

Western Azalea (*Rhododendron occidentale*) (fig. 54). One of the showiest of all the Sierra Nevada shrubs, usually 1–3 m tall, with several stems. The deciduous leaves are thin, 3–8 cm long. Flowers are white to various shades of pale pink, 3.5–5 cm long, and occur from April through July. In the Sierra Nevada the Western Azalea occurs from Kern County north on the western slopes, mainly along streams, in the Yellow Pine Forest, Red Fir Forest, and mixed evergreen forests.

Blueberry or Huckleberry (*Vaccinium*)
Three species of *Vaccinium* occur in the Sierra Nevada.

[93]

TABLE 21. Species in the Genus *Vaccinium*

	Height dm	Calyx	Corolla length dm	Berry color	Berry length dm
V. *nivictum*	0.5–3	not lobed	5–6	blue-black	5–7
V. *occidentale*	3–7	deeply lobed	4	blue-black	6
V. *parvifolium*	10–40	not lobed	4–6	bright red	6–9

All are shrubby, with small alternate leaves, 1–3 cm long. The flowers have urn- or bell-shaped corollas with very small lobes. The fruit is a berry and is situated below the other flower parts. The berries of all three kinds are edible when ripe. Table 21 will help distinguish among the kinds.

Fig. 55. *Vaccinium occidentale*

Sierra Bilberry or Dwarf Huckleberry (*Vaccinium nivictum*). A very low shrubby plant of wet meadows in the Subalpine Fir Forests and Alpine Fell Fields, usually between 8000 and 12,000 feet.

Fig. 56. *Styrax officinalis*

Western Blueberry (*Vaccinium occidentale*) (fig. 55). Occurs in wet meadows and boggy places from Tulare County north in the Lodgepole Pine and Subalpine Fir Forests.

Red Huckleberry (*Vaccinium parvifolium*). The tallest of the three kinds of huckleberry in the Sierra Nevada. It occurs at much lower altitudes than the others, usually below about 5000 feet, from Fresno County north in shaded areas.

In addition to the above, Cranberry (*Vaccinium macrocarpum*) has been collected in a swamp in Nevada County. Otherwise it is known only from Eastern North America. It may represent an introduction into the Sierra Nevada. It has a deeply divided corolla and a red berry 1–2 cm long.

STYRAX FAMILY (STYRACACEAE)
California Styrax (*Styrax officinalis*) (fig. 56). A deciduous shrub, usually 1–3 m tall, with leaves alternate,

Fig. 57. *Eriodictyon californicum*

roundish in outline, and 2–7 cm long. Flowers are numerous, in showy clusters at the tips of the twigs, white, 12–18 mm long. The fruit is globose, not very fleshy, 12–14 mm long.

California Styrax is a shrub of lower elevations below 3000 feet on the western slopes of the Sierra Nevada from Tulare County north. It occurs in Chaparral, Foothills Woodland, and Yellow Pine Forest, usually in open rocky areas.

Olive Family (Oleaceae)

Flowering Ash (*Fraxinus dipetala*) (pl. 6, e). A deciduous shrub or small tree with twigs and young branches 4-angled. The leaves have 3–7 or occasionally 9 leaflets, 2–4 cm long. The flowers are in clusters to 12 cm long; each flower has two small white petals, about 5 mm long; hence the origin of the specific name. The fruit is 2–3 cm long, has a single seed and a long wing, and is adapted for wind dispersal.

Also known as Foothill Ash, this shrub occurs on

[96]

Fig. 58. *Salvia dorrei*

dry slopes in Chaparral and Foothills Woodland, below 3500 feet. Oregon Ash (*Fraxinus latifolia*) also
occurs in the Sierra Nevada, but it is a very large tree
and its flowers lack petals.

PHLOX FAMILY (POLEMONIACEAE)
Granite Gilia (*Leptodactylon pungens*) (pl. 6, *f*). A
low shrub, 1–8 dm tall, with alternate leaves, 8–15 mm
long, which are divided into 3–7 sharp-pointed lobes.
The flowers are 1.5–2.5 cm long, occur singly in the axils of the leaves, and have the petals united to form a
very narrow tubular corolla. Flower color is white with
tinges of pink or purple. Granite Gilia occurs between
about 4000 and 12,000 feet in dry sandy or rocky soils
in a number of plant communities throughout the Sierra
Nevada.

WATERLEAF FAMILY (HYDROPHYLLACEAE)
Yerba Santa (*Eriodictyon californicum*) (fig. 57). An
erect aromatic shrub with stems growing to 3 m in
height. The thick leaves are to 10 cm long; the upper

[97]

surfaces are sticky and shiny, the lower surfaces gray-ish and hairy. Older leaves are blackish due to a sooty fungus, a member of the genus *Heterosporium* (this represents an interesting symbiotic relationship, the de-tails of which are not completely known, in which the fungus in return for a habitat presumably confers some advantage to the Yerba Santa). The funnel-shaped flowers are 1.5–2.5 cm long and occur in clusters at the tips of new growth. Yerba Santa occurs throughout the foothills of the Sierra Nevada below about 5000 feet, mainly in Chaparral.

MINT FAMILY (MENTHACEAE)

Desert Sage, sometimes known as Gray Ball Sage, (*Salvia dorrii*) (fig. 58). A shrubby plant, usually under 1 m tall. Leaves are opposite, 7–16 mm long. Flowers are two-lipped, with the petals united at the base into a tube; flower color is violet-blue; the flowers are in whorls subtended by purplish bracts. Desert Sage is a typical example of the mint family, having opposite leaves, squarish stems, aromatic foliage, and two-lipped corollas. This species occurs at lower elevations along the eastern slopes of the Sierra Nevada.

FIGWORT FAMILY (SCROPHULARIACEAE)

The figwort family, also known as the snapdragon fam-ily, is a large one with many members in California. Most are herbaceous. Some of the beard tongue (*Penstemon*) species considered below are doubtfully shrubs. Occasionally individual *Penstemon* plants in favorable circumstances do become very woody, with substantial branches.

Two genera occur in the Sierra Nevada; *Mimulus* has yellow to salmon-colored flowers; those of *Penstemon* are white to pink or purple. The leaves of *Mimulus* are sticky to the touch while those of *Penstemon* are not. *Penstemon* has a sterile stamen in addition to the four functional ones; *Mimulus* lacks the sterile stamen.

[98]

Fig. 59. *Mimulus aurantiacus*

Bush Monkey Flower (*Mimulus*)

Three species of bush monkey flowers occur in the Sierra Nevada. All are small evergreen shrubs or subshrubs, rarely more than 1.5 m tall. (In some botanical works, the shrubby members of *Mimulus* are placed in a separate genus, *Diplacus*). The leaves are opposite and sticky to the touch. The flowers are various shades of yellow to salmon colored; there is rarely more than one per leaf at the ends of the young twigs. The common name is derived from the fact that some people can see monkey faces in the corollas. Table 22 distinguishes among the three species.

Orange Bush Monkey Flower (*Mimulus aurantiacus*) (fig. 59). Occurs below about 2000 feet from Tulare County north, mainly in Chaparral.

Notch-petaled Bush Monkey Flower (*Mimulus bifidus*). Occurs from Placer County north, from about 1000 to about 5000 feet, in Foothills Woodland and the Yellow Pine Forest.

[99]

TABLE 22. Species in the Genus *Mimulus*

	Corolla color	Corolla length	Flower stalk length	Leaf length
M. aurantiacus	yellow-orange	3.5–4.5 cm	7–15 mm	3–7 cm
M. bifidus	pale yellow	5.5–6.5 cm	8–15 mm	3–4.5 cm
M. longiflorus	salmon	5–6 cm	2–7 mm	4–8 cm

Salmon Bush Monkey Flower (*Mimulus longiflorus*) (pl. 7, *a*). Occurs from Fresno County south below about 5000 feet, in Chaparral and Foothills Woodland.

Beard Tongue or Penstemon (*Penstemon*)
Penstemon is a very large genus, with some 58 species in California alone, but most of them are herbaceous perennials. The leaves are opposite, and the flowers are tubular, two lipped, and generally very showy. Six shrubby Penstemons occur in the Sierra Nevada and may be distinguished with the aid of Table 23; it is interesting to note that five of the six were named after individuals, all either botanists or botanical collectors.

Gaping Penstemon (*Penstemon breviflorus*) (pl. 7, *b*). Quite common in the Sierra Nevada on dry rocky slopes from about 3000 to 8000 feet, in a number of plant communities.

Bridge's Penstemon (*Penstemon bridgesii*) (pl. 7, *c*). Occurs on dry slopes from Kern County north to Alpine County between 5000 and 11,000 feet in several plant communities.

Creeping Penstemon (*Penstemon davidsonii*) (pl. 7, *d*). Occurs in rocky places, often in crevices of rocks, from 9000 to 12,000 in the Subalpine Forests and in Fell Fields, from Tulare County north. This species is sometimes considered to be a variety of *Penstemon menziesii*.

Fig. 60. *Cephalanthus occidentalis*

Bush Beard Tongue (*Penstemon lemmonii*). Ranges from El Dorado County north on wooded slopes from 2000 to 7000 feet.

Mountain Pride (*Penstemon newberryi*) (pl. 7, *e*). Occurs in dry rocky areas from Kern County north from 5000 to 11,000 feet, but is more common at higher elevations.

Rothrock's Penstemon (*Penstemon rothrockii*). Another penstemon of dry rocky areas from 7000 to 10,000 feet, in several plant communities from Mono County south.

Madder Family (Rubiaceae)

Buttonbush (*Cephalanthus occidentalis*) (fig. 60). A shrub or rarely a small tree, with opposite leaves 5–15 cm long. The distinctive feature of this shrub is the dense clustering of the small white flowers into globose heads, 1–2.5 cm in diameter. Fruits resemble small seeds

TABLE 23. Species in the Genus Penstemon

	Height of plant	Corolla color	Corolla length	Leaf characteristics
P. breviflorus	5–20 dm, erect	white, marked with purple lines	15–18 mm	lanceolate, thin, 1–5 cm long, not toothed
P. bridgesii	3–10 dm, erect	scarlet to vermilion	22–35 mm	lanceolate, leathery, 2–5 cm long, not toothed
P. davidsonii	1 dm, matted	purple-violet	18–35 mm	elliptic-round, thick, 0.5–1.5 cm long, not toothed
P. lemmonii	5–15 dm, erect	yellow, marked with purple lines	10–14 mm	ovate-elliptic, thin, 1–6 cm long, toothed
P. newberryi	1.5–3 dm, matted	rose-red	22–30 mm	elliptic-ovate, thick, 1–3.5 cm long, toothed
P. rothrockii	3–6 dm, low, rounded	yellow, marked with purplish lines	10–12 mm	ovate, thin, 0.5–1.5 cm long, toothed or not

TABLE 24. Genera in the Caprifoliaceae Family

Genus	Leaves	Corolla shape	Corolla symetry	Corolla color	Fruit color
Lonicera	1 leaflet	tubular	irregular	yellow or reddish-purple	blue, black, or red
Sambucus	5–9 leaflets	saucer	regular	whitish to pale yellow	blue, black, or red
Symphoricarpos	1 leaflet	bell	regular	white to pinkish	white
Viburnum	1 leaflet	saucer	regular	white	black

Plants	Upper leaves	Flower arrangement in	Berries	Flower color	Flower length
L. caurina	deciduous	not united	pairs	fused	yellow	12–18 mm
L. conjugialis	deciduous	not united	pairs	partly fused	dark purple	6–8 mm
L. hispidula	evergreen	united	whorls	separate	reddish-purple	20–25 mm
L. interrupta	evergreen	united	whorls	separate	yellow	8–10 mm
L. involucrata	deciduous	not united	pairs	separate	yellow	12–15 mm

TABLE 26. Species in the Genus Sambucus

	Plant height	Number of leaflets	Hairs on leaflets	Inflorescence shape	Fruit color
S. mexicana	2–8 m	3–9	glabrous to pubescent	flat-topped	blue-black
S. melanocarpa	1–2 m	usually 5	pubescent, usually glabrous	dome-shaped	black
S. microbotrys	0.5–2 m	7		dome-shaped	red

TABLE 27. Species in the Genus Symphoricarpos

	Growth habit	Corolla color	Corolla length	Corolla lobe length	Berry diameter
S. acutus	trailing	bright pink	4–5 mm	2–3 mm	4– 5 mm
S. parishii	low, sprawling	pink	6–7 mm	3–3.5 mm	6– 8 mm
S. rivularis	erect	white to rose-pink	5–7 mm	2–3 mm	8–12 mm
S. vaccinioides	erect	pink	7–9 mm	2–3 mm	6– 7 mm

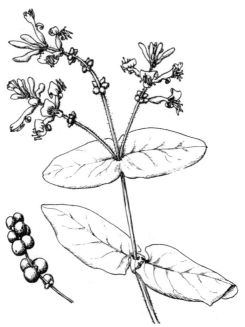

Fig. 61. *Lonicera hispidula*

about 3–4 mm long. Buttonbush occurs along moist banks of streams and ponds below about 3000 feet, from Kern County north in the Foothills Woodland.

Honeysuckle Family (Caprifoliaceae)

Four genera of the honeysuckle family occur natively in the Sierra Nevada. All have opposite leaves. The petals are variously united, and the corolla is situated on top of the ovary. The fruit is edible but in some species does not taste very good. The four genera can be distinguished with the aid of Table 24.

Honeysuckle *(Lonicera)*

Five species of honeysuckles occur in the Sierra Nevada. They are all shrubs or vinelike plants with opposite leaves. The flowers are asymmetrical, the petals united into a tube at the base. The berries are edible, but are not at all tasty. Table 25 will help to distinguish the five kinds.

[104]

Fig. 62. *Lonicera involucrata*

Blue Fly Honeysuckle (*Lonicera caurina*). A low shrub with distinctive blue berries. It occurs from Tulare County north to Nevada County between 5000 and 10,000 feet in moist places in the Red Fir and Subalpine Forests.

Double Honeysuckle (*Lonicera conjugialis*). Has blackish-red berries; rarely over 1.5 m high. It ranges from Tulare County north between about 4000 and 10,000 feet elevation in forest areas.

Hairy Honeysuckle (*Lonicera hispidula*) (fig. 61). Has reddish berries and is distinctively vinelike, often climbing to 3 m. It rarely occurs above about 2500 feet and is found along or near streams on the western slopes in the Foothills Woodland.

Chaparral Honeysuckle (*Lonicera interrupta*) (pl. 7, *f*). Vinelike, often climbing over other plants; red berries. In contrast to the other species, it is found in relatively dry areas in Chaparral, Foothills Woodland, and the Yellow Pine Forest between about 1000 and 6000 feet on the western slopes of the Sierra Nevada.

Black Twinberry (*Lonicera involucrata*) (fig. 62). A shrub, to 3 m in some favored localities, with black berries. It ranges from Tulare County north from sea level to about 10,000 feet, and is usually found in moist areas.

Elderberry *(Sambucus)*

Three species of elderberries occur in the Sierra Nevada. The leaves are compound, that is, composed of several leaflets. The inflorescence is conspicuous, usually white, consisting of many flowers in either flat-topped or dome-shaped clusters. Table 26 distinguishes among the species.

Blue Elderberry (*Sambucus mexicana*). A shrub or small tree in many plant communities up to about 10,000 feet in elevation. *Sambucus caerulea* is here included with *Sambucus mexicana*.

Black Elderberry (*Sambucus melanocarpa*) (pl. 8, *a*). Generally found at higher elevations than Blue Elderberry, 8000 to 12,000 feet, in Lodgepole Pine Forests and Alpine Fields, mainly in the more eastern parts of the Sierra Nevada.

Mountain Red Elderberry (*Sambucus microbotrys*). Common in moist places between about 6000 and 11,000 feet in several forest communities.

Snowberry *(Symphoricarpos)*

The most conspicuous feature of snowberries is the long-persistent, rounded, white berries, each with two seeds. All four species of snowberries in the Sierra Nevada are low, deciduous shrubs, under 1.5 m tall, either erect or somewhat trailing. The leaves on fast-growing

Fig. 63. *Symphoricarpos acutus*

new shoots are often lobed, and larger than those produced later in the season. The flowers are small, bell-shaped, 4–9 mm long, and range in color from white to pink or rose-pink. The plants are browsed by both sheep and cattle.

This group of plants is recognized as a difficult genus by botanists. Flower size and proportions of the free to the united portions of the corolla are important characters. Table 27 will help in distinguishing among the various species.

Creeping Snowberry (*Symphoricarpos acutus*) (fig. 63). Has a creeping or trailing habit. It occurs in damp places from about 3500 to 8000 feet in the forests on the western slope of the Sierra Nevada from Tulare County north.

Parish's Snowberry (*Symphoricarpos parishii*). Occurs on dry rocky slopes and ridges from 4000 to 11,000

TABLE 28. Genera in the Family Asteraceae

Genus	Height	Leaf length	Plants spiny	Ray flowers	Flower color	Number of flowers per head	Male and female flowers	Pappus characteristics
Artemisia	0.1–3 m	0.5–6 cm	in one species	absent	whitish to cream	3–20	on same plant	lacking
Baccharis	1–4 m	1.5–10 cm	no	absent	white to yellowish	very many	on different plants	soft, slender bristles
Brickellia	0.3–1 m	0.7–4 cm	no	absent	white to purplish	8–22	on same plant	barbed or feathery bristles
Chrysothamnus	0.3–2 m	1–8 cm	no	absent	golden yellow	5–11	on same plant	soft, slender bristles
Haplopappus	0.1–5 m	2–6 cm	no	absent or present	golden yellow	4–40	on same plant	slender bristles
Tetradymia	0.2–1.2 m	0.5–1 cm	yes	absent	white	4–6	on same plant	slender bristles

[108]

TABLE 29. Species in the Genus Artemisia

	Height	Spinescent	Flower color	Leaf length	Leaf lobing or toothing
A. cana	0.4–9 m	no		2–6 cm	not lobed, usually no teeth
A. spinescens	0.1–0.4 m	yes		0.5–0.8 cm	5–7 parted
A. tridentata	0.5–3 m	no		1–4 cm	usually with 3 terminal teeth

Fig. 64. *Viburnum ellipticum*

feet from Tulare County north. This species was named for one of the pioneer southern California botanists, Samuel Bonsal Parish.

Common Snowberry (*Symphoricarpos rivularis*) (pl. 8, *b*). Has the largest berries of the group. It is common on moist stream banks on the western slope of the Sierra Nevada usually below about 3000 feet.

Mountain Snowberry (*Symphoricarpos vaccinioides*). Found on relatively dry sites, usually on rocky slopes, from about 5000 to 10,500 feet from Fresno County north.

Western Viburnum (*Viburnum ellipticum*) (fig. 64). A deciduous shrub from 1–4 m tall. Leaves are about 3–7 cm long, with 3–5 conspicuous veins from the top of the leaf stalk. The white flowers occur in clusters at the ends of the small branches and are under 1 cm across. Berries are black, about 1 cm long. Although Western Viburnum is common in Oregon and Washington, it is rare in the Sierra Nevada, being

[109]

TABLE 30. Species in the Genus Baccharis

	Height	Degree of woodiness	Leaf shape and length	Leaf toothing	Leaf stickiness
B. douglasii	1–2 m	somewhat herbaceous above	lanceolate 3–10 cm	not or minutely toothed	glandular not glandular
B. pilularis	1–4 m	very woody	oval, 1.5–2.5 cm	5–9 conspicuous teeth	
B. viminea	2–4 m	willowy	lanceolate, 2.5–9 cm	without or with minute teeth	sometimes sticky

TABLE 31. Species in the Genus Haplopappus

	Height	Leaf shape	Leaf length and width	Twigs	Number of ray flowers	Number of disk flowers
H. arborescens	6–30 dm	filiform	3–6 cm × 2 mm	without hairs, but resinous	0	18–23
H. bloomeri	1.5–5 dm	usually filiform	2–6 cm × 0.5–3 mm	without hairs to somewhat felty	0, or usually 1–5	4–13
H. cuneatus	10–50 dm	ovate to rounded	0.5–2 cm × 3–10 mm	without hairs, resinous	1–5, or usually 0	16–28
H. macronema	1–4 dm	oblong	1–3 cm × 3–6 mm	with a white, felty layer of hairs	0	10–26
H. suffruticosus	2–4 dm	oblanceolate	1–3 cm × 1.5–5 mm	glandular, without hairs	3–6	18–40

TABLE 32. Species in the Genus *Tetradymia*

	Height	Leaf length	Spines	Number of flowers per head	Number of bracts
T. axillaris	6–12 dm	1 cm	conspicuous	6–7	5–6
T. canescens	2–6 dm	2–3 cm	sometimes leaves weakly spine-tipped	4	4–5
T. glabrata	3–9 dm	0.5–1 cm	leaves weakly spine-tipped	4	4–5

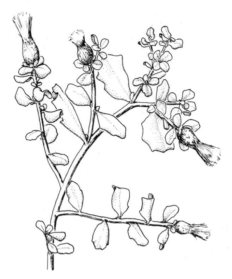

Fig. 65. *Baccaris pilularis*

known only from Fresno and Eldorado counties below about 4500 feet.

SUNFLOWER FAMILY (ASTERACEAE)

The Sunflower Family is common in the Sierra Nevada; some representatives are to be found in every plant community. However, only a relatively few species are shrubs. The distinctive characteristic of this family is the inflorescence, what many mistake to be a single flower: the flowers are closely clustered into heads, with a few to many bracts subtending the flowers. In a typical member of the family, there are two kinds of flowers: the ray flowers are irregular and have elongate petal like corollas; the disc flowers are regular and have generally tubular corollas. The ray flowers, when present, are around the outside of the central mass of disc flowers; in most of the Sierra Nevada shrubs, the ray flowers are lacking. The pappus, thought by some to be a highly modified calyx, is, in the flowers of the Sierran shrubs, either lacking or composed of bristles.

[112]

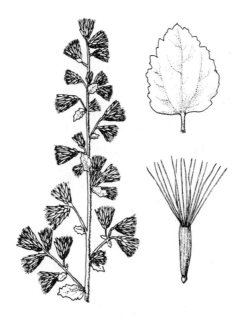

Fig. 66. *Brickellia californica*

The fruit is located below the corolla and other flower parts, contains one seed, and is a dry akene. Table 28 will help in distinguishing among the six genera of Asteraceae present in the Sierra Nevada.

Sagebrush (*Artemisia*)

Three species of shrubby sagebrushes occur in the Sierra Nevada. All are aromatic when the leaves are crushed. Usually the foliage is grayish and hairy. The flowers are in elongate clusters and usually quite small. Table 29 will aid in distinguishing the species.

Hoary Sagebrush (*Artemisia cana*). A small shrubby plant with densely hairy twigs. It ocurs in dry, usually rocky, areas from about 5000 to 10,500 feet from Mono County north on the eastern slopes of the Sierra Nevada.

Bud Sagebrush (*Artemisia spinescens*). Highly branched and with a spinescent habit, due to the lateral

[113]

branches which become spines after they die. This species occurs in dry areas along the eastern slopes of the Sierra Nevada, usually below 5000 feet.

Basin Sagebrush (*Artemisia tridentata*) (pl. 8, *c*). One of the most common shrubs of the Great Basin region of western North America and adjacent regions. It is the most common of the three sagebrushes of the Sierra Nevada. Its specific name, *tridentata*, refers to the easily recognized, three-toothed leaves. Basin Sagebrush occurs in a number of plant communities, particularly the Sagebrush Scrub, throughout the eastern slopes of the Sierra Nevada, and in many places on the western slopes at higher altitudes. This is a variable species, and a number of botanists have recognized additional species within what we are here considering *Artemisia tridentata*.

Coyote Brush *(Baccharis)*
One of the distinctive features of *Baccharis* is that the male and female flowers are on different plants. The inflorescences or heads lack ray flowers. Table 30 gives some of the characters needed to distinguish among the three species in the Sierra Nevada.

Douglas' Baccharis (*Baccharis douglasii*). Woody only at the base. It is quite common below about 1500 feet along streams along the western slopes of the Sierra Nevada from Tulare County north.

Chaparral Broom or Coyote Brush (*Baccharis pilularis*) (fig. 65). A very common California shrub, often acting as a weed, and indicative of overgrazing. It is especially common in the Sierra Nevada below about 2000 feet in Chaparral and Foothills Woodland from Tuolumne County north. The erect Chaparral Broom is usually referred to subspecies *consanguinea* to distinguish it from thet prostrate coastal form.

Mule Fat (*Baccharis viminea*) (pl. 8, *d*). Occurs at low elevations, usually below 1500 feet, in stream beds and other places where the water level fluctuates, along the western slopes of the Sierra Nevada.

Brickellbush *(Brickellia)*
Two species of brickellbushes occur in the Sierra Nevada. Both have triangular-ovate leaves. The flowers lack rays, and the flower heads are arranged in clusters.

California Brickellbush (*Brickellia californica*) (fig. 66). A shrub from 0.5–1 m tall. The leaves are broadly triangular-ovate, 1–4 cm long, coarsely toothed, and covered with short stiff hairs. This species occurs between about 4500 and 8000 feet in the Yellow Pine and Red Fir Forests from Mariposa County north on the slopes of the Sierra Nevada.

Little-leaved Brickellbush (*Brickellia microphylla*). A smaller plant than the preceding, 3–6 dm tall. Leaves are also smaller, usually 0.7–2 cm long, and have longer, less stiff hairs on the surface; leaf margins have few to no teeth. It occurs in dry areas from 3000 to 8000 feet on the eastern slopes of the Sierra Nevada from Mono County north.

Rabbitbrush *(Chrysothamnus)*
Both of the species of rabbitbrushes found in the Sierra Nevada are exceedingly variable, and numerous kinds of subspecies have been described. Rabbitbrush often forms extensive stands on the more arid slopes along the eastern side of the Sierra Nevada. The bright yellow flowers are very distinctive. Only disc flowers are present in *Chrysothamnus*. The clusters of heads in *Chrysothamnus parryi* form a narrow inflorescence. while those of *Chrysothamnus nauseosus* are much wider, rounded to flat-topped.

Common Rabbitbrush (*Chrysothamnus nauseosus*)

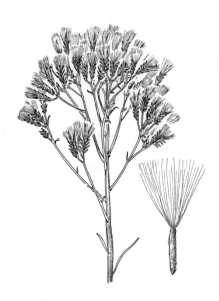

Fig. 67. *Chrysothamnus nauseosus*

(fig. 67). A shrub with several erect branches from the base, generally 3–20 cm tall. Branches are covered with a feltlike layer of hairs; leaves are 2.7 cm long and narrow; flowers are 7–21 mm long, golden-yellow, usually five to one head, arranged in rounded to flat-topped clusters. Common Rabbitbrush occurs throughout the Sierra Nevada below about 8000 feet in open areas in a number of plant communities.

Parry's Rabbitbrush (*Chrysothamnus parryi*). About 3–5 cm tall, the branches covered with a dense growth of hairs giving them a feltlike texture. Leaves are 1–8 cm long and generally rather narrow. Flowers are golden-yellow, 8–11 mm long, usually 5–11 per head. This shrub occurs widely in open areas in the Sierra Nevada from about 3000 to 11,500 feet.

Goldenbush *(Haplopappus)*
This is a large group of yellow-flowered herbs and small

shrubs. Five species occur in the Sierra Nevada. Some have both ray and disc flowers, others have only disc flowers. Table 31 should aid in identification of individual species of goldenbush.

Golden Fleece (*Haplopappus arborescens*). An erect shrub with aromatic leaves. It is usually found in Chaparral and Foothills Woodland from Tulare to Nevada counties between 300 and 4000 feet, but in the southern Sierra Nevada it occurs occasionally at much higher altitudes.

Bloomer's Goldenbush (*Haplopappus bloomeri*) (pl. 8, *e*). A low compact shrub of rocky or sandy areas from about 3500 to 10,000 feet from Tulare County north in several plant communities.

Wedgeleaf Goldenbush (*Haplopappus cuneatus*). Easily identified by its broad leathery leaves. It is a low-growing shrub occurring on rocky slopes and in crevices of granitic rocks from about 3000 to 9000 feet from Tulare County north.

Whitestem Goldenbush (*Haplopappus macronema*). Easily distinguished by the dense layer of white felty hairs on the twigs. It occurs at high elevations, 9000 to 12,000 feet, from Tulare to Nevada counties, often above timberline.

Singlehead Goldenbush (*Haplopappus suffruticossus*). Has few flower heads at the tips of the branches. It occurs at higher elevations, 8000 to 12,000 feet, on open rocky slopes from Tulare County north to Nevada County.

Horsebrush *(Tetradymia)*

Three species of Horsebrush occur on the eastern slopes of the Sierra Nevada. All are small shrubs with woolly twigs, at least when young, and white flowers. The flowers are few per head, as are the bracts subtending the flowers. Table 32 will help in identification.

Cotton Thorn (*Tetradymia axillaris*) (pl. 8, *f*). Easily recognized by its spines which are modified leaves. It occurs on dry slopes between about 3000 and 6500 feet on the eastern slopes of the Sierra Nevada north to Mono County.

Spineless Horsebrush (*Tetradymia canescens*). Leaves are 1–4 mm wide and generally not very spiny. It occurs between about 4000 and 10,000 feet on dry slopes in several plant communities.

Littleleaf Horsebrush (*Tetradymia glabrata*). Leaves are about 1 mm wide and spinescent. It occurs about 7000 feet along the eastern slopes of the Sierra Nevada in several plant communities.

GLOSSARY OF TECHNICAL TERMS

Akene: A dry 1-seeded indehiscent fruit.

Alternate: Placed singly at different levels on an axis; usually with respect to leaves which are neither opposite nor whorled.

Anther: The portion of the stamen which contains pollen.

Bract: A small leaflike structure usually subtending a flower or a portion of an inflorescence.

Bur: A spiny fruit.

Calyx: A collective term for the outermost whorl of the perianth, composed of sepals which may be distinct or fused to form a synsepalous calyx, usually green, in some flowers colored or completely absent.

Capsule: A dry dehiscent fruit containing two or more carpels.

Catkin: A small scaly spike or spikelike inflorescence usually flexuous and pendulous.

Compound leaf: A leaf composed of two or more distinct blades or leaflets.

Deciduous: Not persistent; woody plants that lose their leaves in the fall are said to be deciduous.

Disc flower: The tubular radially symmetrical flower in the center of the head of members of the sunflower family.

Elliptic: A shape in the form of a flattened circle that is more than twice as long as broad.

Entire: A margin that is smooth, lacking teeth or incisions.

Filament: The stalk of the stamen which supports the anther.

Filiform: Threadlike.

Hypanthium (floral tube): An enlarged cuplike structure below the calyx.

Indehiscent: Not splitting open.

Inflorescence: The arrangement of flowers on an axis or a series of axes.

Irregular: Usually said of flowers that are bilaterally symmetrical.

Lanceolate: Lance-shaped with the broader end toward the base.

Linear: Long and narrow with parallel margins.

Node: The part of the stem which bears a leaf.

Opposite: Placed at the same level on an axis and opposite each anther; one part in front of another part.

Ovary: The portion of the pistil containing the ovules (which become seeds after fertilization).

Ovule: The body in an ovary which becomes a seed.

Pappus: The modified calyx in the sunflower family, consisting variously of scales, hairs, bristles, or absent.

Petal: One unit of the corolla, usually colored.

Pistil: The ovule-bearing organ of a flower consisting of stigma and ovary, usually with a style between.

Ray flower: The bilaterally symmetrical flower of the sunflower family.

Receptacle: The portion of an axis that bears the flower parts; the expanded portion of the axis that bears the flowers in the sunflower family.

Regular: Usually said of flowers which are radially symmetrical.

Sepal: One unit of the calyx, usually green.

Simple: Of one part.

Stamen: The male part of the flower consisting of flament, connective and anther.

Stigma: The part of the pistil that receives pollen.

Style: The portion of the pistil between the stigma and the ovary.

Whorl: Several structures arranged in a circle about an axis.

REFERENCES

Abrams, L., and R. S. Ferris. *Illustrated Flora of the Pacific States.* Stanford University Press. 1923–1960.

Jepson, W. L. *A Manual of the Flowering Plants of California.* Associated Students Store, Berkeley. 1923–1925.

Johnston, V. R. *Sierra Nevada.* Houghton Mifflin Co., Boston. 1970.

McMinn, H. E. *An Illustrated Manual of California Shrubs.* J. W. Stacey, Inc., San Francisco. 1939.

Munz, P. A. *A California Flora and Supplement.* University of California Press, Berkeley and Los Angeles. 1973.

Peterson, P. V., and P. V. Peterson, Jr. *Native Trees of the Sierra Nevada.* University of California Press, Berkeley and Los Angeles. In press.

Storer, T. I., and R. L. Usinger. *Sierra Nevada Natural History.* University of California Press, Berkeley and Los Angeles. 1963.

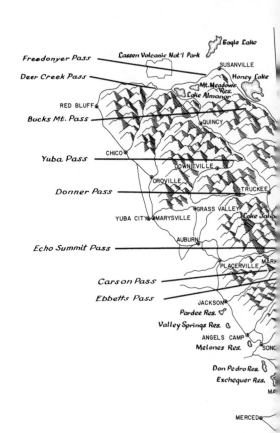

MAP OF THE SIERRA NEVADA

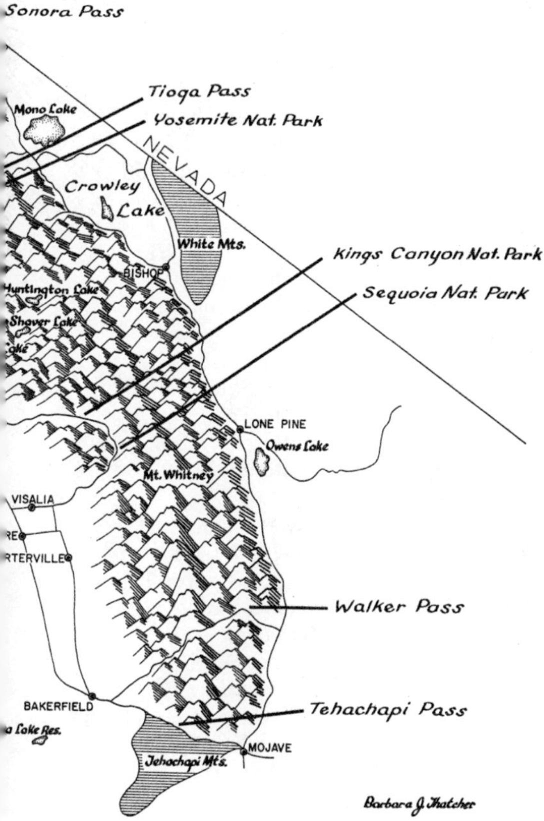

Sonora Pass

Tioga Pass

Mono Lake

Yosemite Nat. Park

NEVADA

Crowley
Lake

White Mts.

Kings Canyon Nat. Park

BISHOP

Sequoia Nat. Park

Huntington Lake

Shaver Lake

LONE PINE

Owens Lake

Mt. Whitney

VISALIA

RE

RTERVILLE

Walker Pass

BAKERFIELD

Tehachapi Pass

a Lake Res.

MOJAVE

Tehachapi Mts.

Barbara J. Thatcher

INDEX